AMAZON TO ZODIAC

101 SCIENTIFIC CURIOSITIES ACROSS TIME AND SPACE

BARRY EVANS

Prometheus Books
Essex, Connecticut

Prometheus Books
An imprint of The Globe Pequot Publishing Group, Inc.
64 South Main St.
Essex, CT 06426
www.globepequot.com

British Library Cataloguing in Publication Information available

Library of Congress Cataloging-in-Publication Data available
ISBN 978-1-4930-9834-7 (cloth)
ISBN 978-1-4930-9558-2 (paper)
ISBN 978-1-4930-9561-2 (ebook)

CONTENTS

CONTENTS

INTRODUCTION

As an unabashed lover of science, my goal in writing this book of essays is to demystify inaccessible, remote topics; stimulate your curiosity; and encourage you to marvel at the amazing world around and within us.

We're a novelty-seeking species, primed to ask questions when looking up at the stars on a clear night or tasting a new fruit for the first time. What is this? How did it come to happen? Why is it the way it is? Science helps us answer such questions, but it's not just a fact-spewing machine; it offers us much more. It helps us frame and reframe our sense of ourselves and of our species and, through our heightened understanding, make better decisions. Best of all, it arouses our natural sense of wonder.

Each of these essays tackles and clarifies a different aspect of science in our complex world, whether the topic is Mastodons in Greenland, Vlad the Impaler, Roman dildos, crystal sets, or Ötzi the Iceman. And as I've tried to convey, science can be funny too.

Because of my very hierarchical, top-down education, it wasn't until I was an adult that I came to appreciate science, and I know from firsthand experience that a little goes a long way. This is why most articles are 5- to 10-minute reads—perfect for reading on a commute, waiting your turn at the DMV, or in the bathroom. And because pictures can explain better than text, most essays are illustrated.

I hope you'll be informed and entertained by the myriad scientific curiosities here. Enjoy!

Barry Evans
Eureka, California
barryevans9@yahoo.com
March 2026

ANTHROPOLOGY

Where we came from

1

A IS FOR AMAZON

Not all the sistahs picked berries

Their marriage law lays it down, that no girl shall wed until she has killed a man in battle.

—Herodotus, ca. 600 BC, writing about the Sauromatae, mythical descendants of Amazons

The world has known plenty of powerful female leaders, from Queen Dido (Carthage), Cleopatra (Egypt), Boadicea (of the Iceni, the British tribe that nearly defeated the Roman invaders), Elizabeth I (England), Catherine the Great (Russia), Angela Merkel (Germany), Claudia Sheinbaum (Mexico), and many others. Yet in the United States, the idea of women leaders—whether in politics or in business—is still seen as revolutionary! Such patriarchal thinking goes back to the old, stubbornly ingrained attitude of (white male) anthropologists of the recent past who presented ancient tribal societies as a dichotomy: male hunters and female gatherers. Read, tough dudes and wimpy women. While there is some contemporary truth in this—the roles of modern-day hunter-gatherer people like Tanzania's Hadza and southern Africa's San are firmly gender defined—the past few decades have unearthed plenty of counterexamples.

For instance, in 2018, anthropologists from the University of California, Davis, discovered the skeleton of a woman who was buried with

Imaginary scene high in the Andes 9,000 years ago. MATTHEW VERDOLIVO/UC DAVIS IET ACADEMIC TECHNOLOGY SERVICES, USED WITH PERMISSION

her full kit of hunting tools, including dozens of projectile points, at a site in the Peruvian Andes known as Wilamaya Patjxa. (The hunter's sex was determined by a relatively new technique in which tooth enamel can be shown to come from a male or a female by analyzing a protein called amelogenin.)

The fact that there were women hunters in that society is no big deal, according to archaeologist Bonnie Pitblado from the University of Oklahoma: "Living high up in the Andes at 13,000 feet, if you can do that, surely you can bring down a deer." The discovery led archaeologists to recheck some 400 burials in the Americas dated to be more than 8,000 years old. Of the 27 skeletons found buried with big-game hunting tools, 11 were female, leading to the statistical conclusion that up to half the hunters of that time may have been female.

Back in the Old World, in the early 1990s, a joint team of U.S. and Russian archaeologists was investigating Indo-European–speaking herders, so-called Scythian-Sarmatians, who lived on the steppes of the southern Urals some 2,300 years ago. While excavating *kurgans* (tumulus burial mounds), they found the remains of several females who had been buried with their weaponry: swords, daggers, and—in one case—a quiver holding 40 bronze-tipped arrows. From her bowed leg bones, this owner seems to have spent much of her life on horseback.

Another Old World site, on an island west of Stockholm, Sweden, yielded the 1,000-year-old skeleton of a Viking along with a sword, an axe, a spear, arrows, a battle knife, and two shields when it was excavated in 1889. Long thought to be remains of a male, bone and DNA analyses recently showed that the "Birka Viking warrior" was female.

Okay, this doesn't automatically prove that there were legions of female fighters in what's known to have been a thoroughly patriarchal Viking society, and there's plenty of controversy about what this finding means. As *New York Times* contributor Annalee Newitz pointed out, "One female warrior does not mean that many women were leaders, just as the reign of Queen Elizabeth I was not part of a larger feminist movement." Still, increasingly, archaeologists and anthropologists are finding evidence for a more nuanced view of ancient gender roles than previously thought and that hunting and fighting weren't confined to men.

AMAZONS

Any discussion of female warriors has to reference Amazons. Herodotus, the so-called Father of History, had much to say about them. According to him, the legend is that Amazons were a race of women warriors who nearly bested Greek soldiers while trying to free one of their own during the Attic War.

He claimed that the Amazonian capital Themiscyra was located in present-day northern Turkey near the shore of the Black Sea. Brooking no male presence, the women took care of procreation in an annual get-together with the tribe who lived on the other side of the mountain, the Gargareans. Baby girls were brought up to become Amazon warriors, while baby boys were sent back to their fathers. When they weren't reproducing, according to Herodotus, the Amazons spent their days either pillaging their neighbors or founding new cities, including Ephesus and Smyrna.

As background, Herodotus was elaborating on Homer's earlier tale of the Trojan War, the *Iliad*, which had its oral origins in the Late Bronze Age (ca. 1200 BC). Homer's Amazons were the "equal of men"

Section of the "Amazons versus Greeks" marble frieze that once decorated the Mausoleum of Halikarnassos (modern Bodrum, Turkey), sculpted about 350 BC. All the Amazons are depicted with both breasts intact. The frieze is now in the British Museum next to statues of the guy for whom the mausoleum was built (and from whom we get the word), Satrap (King) Mausolus of Caria and his wife. AUTHOR

(*antianeira*), and he began a literary tradition of the women allying themselves with the (losing) Trojan side. A later Greek myth has the hero Hercules stealing Amazon queen Hippolyta's magic girdle as his ninth "labor," while in another version, Hercules abducts the queen, giving her to Theseus, the Greek hero who supposedly founded Athens.

What could have inspired these legends about warrior women who could inspire fear and awe in the male-dominated Athenian *polis*? One theory links the legendary Amazons with those real Scythian-Sarmatian women warriors mentioned earlier.

A classical scholar writing in the nineteenth century, Johann Bachofen, went so far as to propose that human societies began as matriarchies, only switching to patriarchies with the rise of agriculture. The story goes that composer Richard Wagner was a fan of Bachofen: in his

Ring Cycle, Brünnhilde and her Valkyries were the Nordic equivalent of Sythian Amazons.

The very name "Amazon" is also controversial. The ancient Greeks had a folk etymology that the word means "without breast" (*a-mazos*), in line with a tradition that the Amazons cut, or cauterized, their right breasts in order to better wield a sword or draw a bow. There are other, more convincing etymologies (for example, *ama-zoonais* = "with girdles"). The fact is that Amazons are always shown in their many Greek representations with both breasts intact. While there's uncertainty about the word's origin, we do know that the Amazon River in South America was named by the Spanish explorer Francisco de Orellana, who encountered—and fought—a tribe of warlike women on a tributary of the river in 1542.

Stories of kickass women live on today in comics and movies: Wonder Woman, Lara Croft, Xena, Buffy Summers, and the baddest of them all, Uma Thurman's character in Tarantino's *Kill Bill*. Now there's an Amazon for our times.

2

THE COOKING APE

How fire made us human

Ape. That's what you'd have thought if you met one of our ancestors 2 million years ago: short stature, low brow, massive jaw with a mouth housing huge teeth, small brain, flared ribs enclosing a big belly, slouching gait, and tree-climbing anatomy—all the hallmarks of a bipedal ape. Fast-forward to 1.7 million years ago, put your great-great-great-grandfather into an off-the-rack Brooks Brothers suit (which would fit him just fine), and you'd think "human." What happened during that comparatively brief 300,000 years during which our forebears morphed from chimpanzee-like "habilines" (*Australopithecus habilis* and *Homo habilis*) to human-like *Homo erectus*? Richard Wrangham, professor of biological anthropology at Harvard, has a one-word answer: *cooking*. It's all in his 2009 book *Catching Fire: How Cooking Made Us Human*.

Wrangham's thesis is that around 1.9 million years ago, during the Early Pleistocene, our ancestors learned to control fire, which until then had been a hit-and-miss event. Lightning-initiated fires on the African savanna were irregular phenomena, but at some point, our habiline forebears figured out how to start a fire (probably with sparks produced when they struck pyrite rocks with flint). Fires, even without cooking, would have had an immediate impact on their lives, allowing them to sleep on the ground, for example. Until then, they'd have slept in trees to avoid predators, as most apes do today. But when one of them accidentally

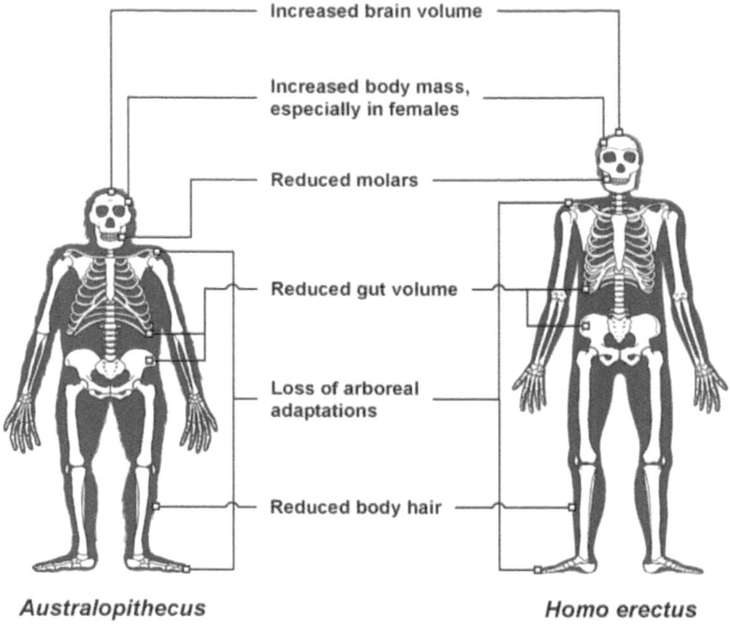

Cooking may have driven the evolution of human-looking *Homo erectus* from ape-like *Australopithecus*. WRANGHAM-CARMODY, USED WITH PERMISSION

dropped a slab of meat into a fire—the first barbecue—the transition from ape to human really got underway.

Cooked meat is essentially predigested meat since heating "denatures" protein, allowing stomach enzymes easier access to animal food. Gelatinizing starch by heating it also helps digestion. Although present-day humans *can* live on uncooked food—just—our bodies are thoroughly adapted to a cooked diet. For example, our chewing teeth (molars) are small compared with all other primates when adjusted for body size. Similarly, our mouths are tiny, our jaw muscles weak, our stomachs small, and our colons short: all the result of cooking, according to Wrangham. He points out that the great apes have to eat twice as much as humans because one-third of their food is indigestible fiber that gets excreted, whereas in humans, less than 10 percent of our food is indigestible.

Other than raw foodists, today everyone, everywhere, cooks their food—and for good reason: *energy*. It takes much more energy to digest food (plant or animal) that is (1) hard, tough, and dense; (2) cold; and (3) in large chunks, like meat cut from an animal. (As it is, our metabolic rate rises about 25 percent after eating.)

But cooking doesn't just give us more energy; it gives us more time. Other primates typically spend most of their waking hours chewing. Wrangham estimates that, absent cooking, humans would be spending about six hours a day just chewing (and the rest of the day hunting and gathering). As it is, most of us spend less than an hour a day actually eating.

FROM APE TO HUMAN

Once our forebears no longer needed all that high-powered, energy-intensive machinery for mastication and digestion, our bodies "humanized." Teeth, jaws, mouths, stomachs, and colons got smaller, while our brains capitalized on the surplus energy made available by cooking. Brains are calorie guzzlers, accounting for 20 percent of the calories we burn while weighing just 2 percent of our total body weight. Those extra calories grew brains. The cranial capacity of (raw food) habilines was about 37 cubic inches and that of (cooked food) *Homo erectus* around 53 cubic inches. Meanwhile, body weight nearly doubled, from about 75 to 135 pounds.

So that's the theory: cooked food led to shrinking everything except our brains, which eventually doubled in size to our present average of 73 cubic inches. According to this scenario, once our ancestors tamed fire and invented cooking some 1.9 million years ago, human evolution took off, beginning with almost-modern-looking *Homo erectus*. Today, our bodies are thoroughly adapted to cooked food.

Once our ancestors no longer needed to chew for hours, they had time to kill, literally. The men (mostly) went hunting for hours or even days, in contrast to, say, chimpanzees, which revert to low-calorie plant food if they haven't killed in the first half hour. In addition to lengthy hunting expeditions, cooking led to huge social changes. Most animals, including our simian cousins, wander around alone and eat where the food is. In contrast, when humans started cooking, they brought the food to the fire, requiring them to sit around and cooperate—or at least not fight, as chimps do as a matter of habit. To successfully socialize in a community, human temperaments had to grow calmer.

GENDER ROLES

Another significant effect, according to Wrangham, was the establishment of gender roles. For the most part, while men hunted, women foraged and cooked. What was to stop a hungry man from stealing food from a woman? *Pair bonding*. A woman needed a man to protect her and her offspring, while a man needed a woman to cook for him at the end of the day. Thus was born the modern family structure.

Not that the fire-and-cooking combo brought only light and love. Fires probably led to the emergence of many infectious diseases, which evolve and spread when people spend time in close contact. Tuberculosis, for example, which has probably killed a billion people over the millennia, still claims over a million lives every year globally.

While the "cooking made us human" theory has plenty going for it, one major piece of the puzzle is missing: archaeological evidence that humans controlled fire 1.9 million years ago. Most researchers agree that we have good evidence for hearths in the form of burnt wood ashes associated with human habitation, going back at least 250,000 years. Beyond that is where the controversy begins, for example, with a recent claim for 1.6-million-year-old fireplaces in Kenya, according to potassium-argon analysis.

Apart from the lack of undisputed evidence for controlled fire 1.9 million years ago, other anthropologists maintain that the critical step in our evolution was when we started eating raw meat, perhaps a million years earlier. (We have evidence of 2.5-million-year-old cut marks on large animal bones.) However, the really profound anatomical changes all began no earlier than when *Homo erectus* emerged, nearly 2 million years ago. The verdict is still out, but right now, Wrangham's "cooking" claim appears a worthy competitor in the several current theories for how and why apes evolved into humans.

3

BY THE SKIN OF OUR TEETH

Our species survived a bottleneck. Just.

When a species gets down to a few hundred individuals, it is considered endangered. This was probably the case with our own species, *Homo sapiens*, whose numbers fell precipitously around 150,000 to 130,000 years ago during the peak of one of the severest ice ages, the "Marine Isotope Stage 6." At that time, all our ancestors lived in Africa when almost the entire continent was arid and uninhabitable. To survive the extreme climate, they had to be in a place where consistent sources of protein and carbohydrate were available. Anthropologist Curtis Marean of Arizona State University believes that Pinnacle Point, a promontory at the very southern tip of South Africa, was that life-supporting sanctuary.

In the 1990s, prior to Marean's investigations, geneticists studying human DNA—especially that found in mitochondria (which is passed on only by mothers)—began to suspect that all 8 billion of us alive today are descended from a tiny African population. Relative to most other species, we have a very low genetic diversity, suggesting that our species passed through a "bottleneck" before surging in numbers and subsequently spreading out of Africa into the rest the world. Some—perhaps all—of our ancestors who made it through the climate-created bottleneck may have done so at Pinnacle Point.

Pinnacle Point is at the center of a particularly benign area where both carbohydrate-rich tubers and bulbs and protein-rich shellfish were

Cave 13B, Pinnacle Point. SACP4, ARIZONA STATE UNIVERSITY, USED WITH PERMISSION

available during a period when the rest of Africa was suffering a climate catastrophe. In particular, the convergence of two ocean currents at the southern tip of Africa encourages shellfish growth, especially during low water temperatures.

Cave 13B, located in a cliff now overlooking the Indian Ocean, is the most significant site found to date. Here, researchers uncovered evidence of early human consumption of shellfish and other marine resources during the Middle Pleistocene, which ended 126 million years ago. During the ice ages, the cave would have been some distance inland, suggesting that in order to make worthwhile treks to the ocean at low tide—when shellfish were available—the cave dwellers would have had a sense of lunar phases (which control the tides). Prior to the Pinnacle Point discoveries, such sophistication was assumed to have arisen much later in our ancestors.

It may well be that this food-rich oasis discovered by early humans was responsible for the survival of our species, and it was there that we made it through the worst period of glaciation—by the skin of our teeth.

More at https://www.scientificamerican.com/article/when-the-sea-saved-humanity-2012-12-07.

4

THE NEANDERTHAL CONNECTION

Telltale genes

Calling some obnoxious dude a "Neanderthal" no longer has the same insult value as it once had since German researchers discovered, about two decades ago, that most of us non-Africans carry between 1 and 4 percent Neanderthal DNA in our genes. This wasn't news to the many archaeologists who have been arguing for some time that fossils show evidence of interbreeding and that Neanderthals and humans lived side by side in excavated caves in, for example, present-day France and Israel. This finding calls into question the traditional designation of Neanderthals as a separate species from extant humans since the ability to interbreed is one of the usual criteria for defining "species."

The discovery was science at its best. Apparently, a consortium of genome researchers came into the project with a very strong bias *against* finding Neanderthal DNA in modern humans, but the results, from several independent lines of evidence, convinced them otherwise. Working with DNA from bone fragments of three Neanderthal women who lived in a cave in Croatia about 40,000 years ago, they were able to reconstruct about 60 percent of the Neanderthal genome. This astonishing feat took five years of intricate analysis using cutting-edge gene sequencing technology. "We used half a gram of bones to produce the three billion base pairs," said Svante Pääbo of the Max Plank Institute for Evolutionary Anthropology, leader of the project. He was awarded the 2022 Nobel Prize in Physiology and Medicine for his work.

Reconstruction of a Neanderthal child from a skeleton found in Gibraltar in 1926. Tomographic scanning was used to convert the remains into a computer model from which a physical model was constructed using stereolithography. Soft tissue was then extrapolated using a technique known as thin plate splining. ANTHROPOLOGICAL INSTITUTE, UNIVERSITY OF ZÜRICH, USED WITH PERMISSION

The researchers compared the newly sequenced Neanderthal DNA with genomes from five present-day humans from China, Papua New Guinea, France, and a San and a Yoruba from southern and eastern Africa, respectively. While they found 1 to 4 percent Neanderthal ancestry in the first three (from Asia and Europe), they found no trace in either of the Africans. This suggests that interbreeding occurred, perhaps in the Middle East, after humans had migrated from Africa around 100,000 years ago but before they evolved into distinct groups in Europe and Asia about 50,000 years ago.

The last Neanderthal died about 30,000 years ago, probably in southern Europe. Except, welcome to your family tree. On average, Neanderthal DNA occupies 2 percent of the genomes of people not of African descent. They are us, our closest human relatives.

5

ALTRUISM'S HIDDEN BENEFITS

Being kind pays off

I'm an easy mark. If someone in our local downtown wants a dollar for a cup of coffee, I instinctively pull out my wallet, even though I'm pretty sure that coffee isn't my petitioner's beverage of choice. Seems that *altruism*—concern for others without the expectation of a reciprocal payoff—is quietly at work in my brain.

When anti-evolutionists attack the notion of natural selection (the process whereby organisms best adapted to their environment are favored to transmit their genes to succeeding generations), they frequently invoke the altruism argument. By definition, altruistic acts exact a cost to the benefactor, meaning that "selfish" genes should outcompete "altruistic" genes. Therefore, it must take a force acting outside of natural selection—God?—to make us act against our own best interests.

Evolutionary psychology offers several responses, none of which require divine intervention. One is *kin selection*. Our brains were evolving when our ancestors hunted and gathered in groups of perhaps 50 to 100 people and when fellow travelers tended to be blood relatives. Evolution would have favored a gene for being kind to one's kin when the net gain to the group exceeded the loss suffered by the "good guy."

A second response is *reciprocal altruism*. Picking the nits out of someone else's hair costs you in the short run (that time might be better spent cooking or feeding yourself), but you'll benefit later when the nit-picking is reciprocated. Like washing your hands, both parties are better off from the exchange of favors.

Beyond kin selection and reciprocal altruism, it's easy to see other reasons why generous genes might sometimes outcompete selfish genes. For example, a man with a reputation for being kind attracts mates—if you're a fertile woman, you want the father of your children to be generous. Also, natural selection should favor behavior that allows us to get along with others—if you have a reputation for generosity, you're less likely to be challenged to a potentially lethal fight than, say, an alpha male is.

It's humbling to realize that so much of our behavior is ruled, out of sight and mind, by our genetic makeup. Much as I'd like to think of myself as a fine, generous person, I know it's really my genes that hand over the money. They just don't understand that you can't get a decent cuppa joe for a buck these days.

6

THE EARLIEST AMERICANS

From Siberia to the Americas

In the early 1970s, the question of when humans first came to the Americas had been decisively answered: 13,500 years ago, at a time when sea levels were 200 feet lower than now, big-game hunters migrated east from Asia across a continent-size land bridge ("Beringia") connecting Siberia to Alaska, then south through an ice-free tundra corridor along the course of the Mackenzie River to Alberta and beyond. This, in a nutshell, is the "Clovis First" model, named for an archaeological site near Clovis, New Mexico, that, starting in 1929, yielded hundreds of distinctive "Clovis" arrowheads and other stone tools made by some of the earliest newcomers to the New World.

The first major crack in the Clovis model came in the mid-1970s, when Vanderbilt archaeologist Tom Dillehay dated campfires at a creek-bank site, "Monte Verde," in southern Chile at least 1,000 years earlier than the Clovis arrivals. (Carbon-14 dating pinpoints when organic material, such as wood, last metabolized, in this case, when it was used as fuel.) Since the discovery of Monte Verde, another dozen or so sites in North and South America have challenged the conventional wisdom of Clovis First, but none yielded cast-iron proof that it was wrong.

The picture changed in 2008, when DNA from human coprolites—fossilized poop—found in the Paisley Caves, 100 miles northeast of

Klamath Falls, Oregon, was found to be up to 14,300 years old. Two years later, a team led by Texas A&M University archaeologist Michael Waters struck gold on the banks of Buttermilk Creek in Texas, with the discovery of some 20,000 stone tools from about 15,500 years ago. (They couldn't date the tools themselves, but a fairly new technique, known as optically stimulated luminescence, allowed them to estimate the date when the surrounding soil last saw sunlight.)

WHITE SANDS

Yet more evidence for a *much* earlier arrival came in 2010, with the announcement of the next best thing to human remains: human footprints. The journal *Science* published a paper in which the 15 authors contend that human footprints found in eight layers of exposed outcrops in New Mexico's White Sands National Park were made on a lakeshore as early as 23,000 years ago. The footprints are undoubtedly human. No other creature leaves the prints like those shown in the photo, probably left by young children and teenagers. Another set of prints seems to show a mother walking with a young child before she picks up and carries the youngster, first on one side of her body, then the other side.

Today, the park is known for its white gypsum dunes stretching for hundreds of square miles, but during a warm period in the middle of the last ice age, the area was a grassy plain with a lake the size of Rhode Island. The tracks from different strata show that humans lived alongside the lake for at least 2,000 years, probably for the hunting opportunities. In addition to giant sloths, researchers have found tracks of ancient camels, mammoths, dire wolves, and saber-toothed cats.

So what's the unequivocal evidence since the prints themselves, first found in 2009, can't be directly dated? The excavators found hundreds of seeds of *Ruppia cirrhosa*, better known as spiral ditchgrass, above, below, and in some cases embedded in the footprints. Using carbon dating, the seeds have been shown to range between 21,000 and 23,000 years old, with the oldest ones lowest, as would be expected in the absence of any disturbance or contamination.

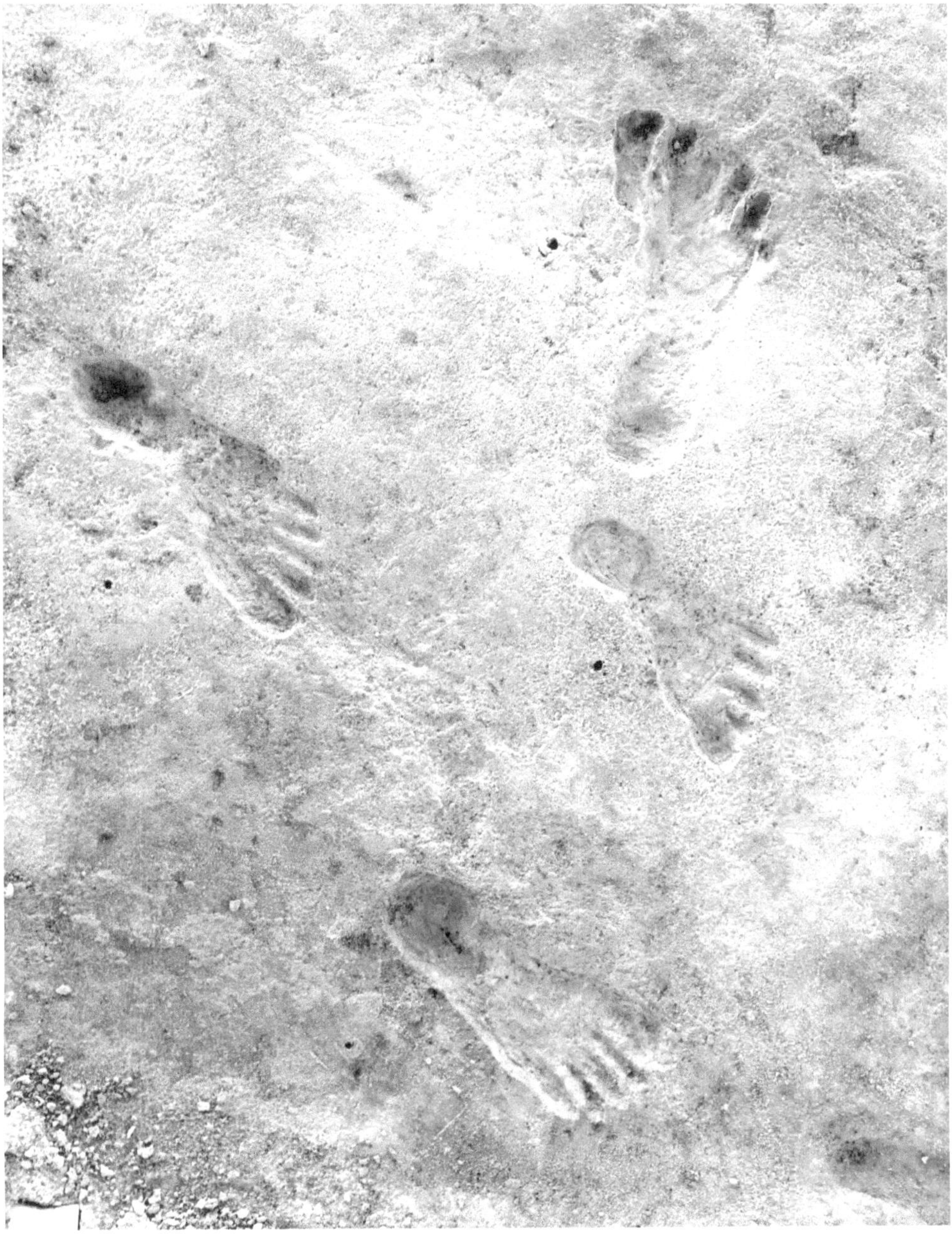

Fossilized footprints of children and teenagers from more than 20,000 years ago, found in White Sands National Park, New Mexico. NATIONAL PARK SERVICE, USED WITH PERMISSION

KELP HIGHWAY

With the Clovis First model pretty well discredited, the next question is, how did those first Americans get here? While most archaeologists still favor the land passage across Beringia and down a wide ice-free corridor (following the opening up of a contiguous ice sheet across present-day Canada), some support the idea of a sea route, with adventurous souls paddling a "kelp highway" along the southern coast of Beringia and down the western shore of the Americas. It was certainly technically feasible—humans first crossed the Timor Sea to Australia 45,000 years ago—and they could have sustained themselves on shellfish and salmon once ice cleared from the coast of British Columbia around 16,000 years ago.

One intriguing piece of evidence demonstrating the seafaring ability of Paleo-Indian sailors comes from Santa Rosa, one of California's Channel Islands. Discoveries starting in 2008 at site "CA-SRI-512" show that

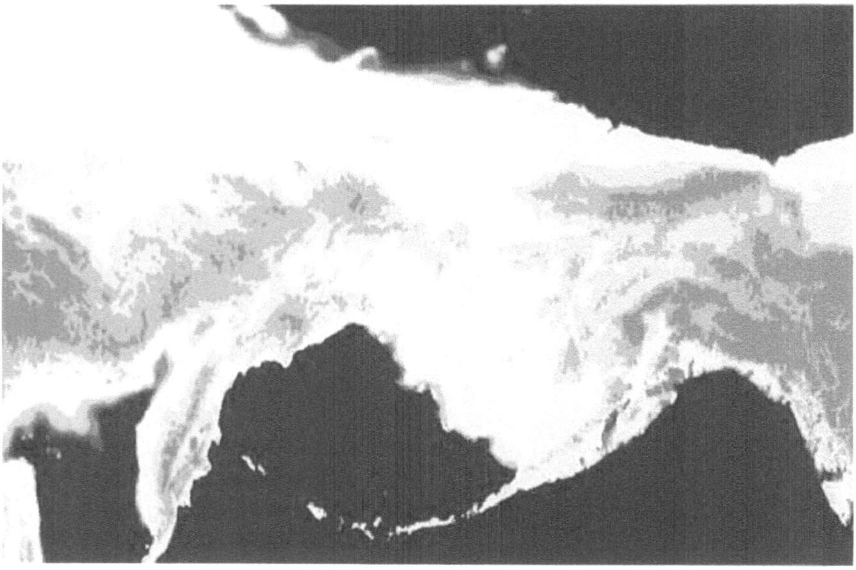

"Beringia" 14,000 years ago. The first Americans may have walked across it or paddled along its southern shore, thence down the west coast of the Americas. (NOAA) For an animated version showing the changing shoreline from 20,000 years ago to the present, check out https://commons.wikimedia.org/wiki/File:Beringia_land_bridge-noaagov.gif.

humans were using stone tools there, similar to ones found in East Asia, to hunt geese, cormorants, and marine mammals at least 12,000 years ago. But did they paddle all the way down the west coast to Chile? That's probably a question that can be answered only by marine archaeologists since any intermediate shoreline habitation sites would be under a couple hundred feet of water. Whichever way they came, those early trailblazers were tough as nails, with a supreme ability to adapt to changing conditions before fanning out across the vastness of the Americas.

So that's the new picture that's slowly gaining credibility: the earliest Americans paddled along the southern coast of Alaska and down the Pacific coast rather than walking across Beringia and migrated as far as present-day New Mexico more than 20,000 years ago. Now if only we could find the bones of the kids who left those footprints, especially if those bones contained recoverable DNA!

The story of the earliest Americans is far from over.

7

AGRICULTURE: OUR WORST MISTAKE?

The Neolithic Revolution

The downturn in human nutrition began long before Ray Kroc brought Big Macs to the masses, Coca-Cola conned Third World countries into buying their addictive sugary concoctions, or Nestlé weaseled its way into infants' stomachs with formula milk. Long, long before.

Archaeologists call it the "Neolithic Revolution," the series of events that started in the Middle East's Fertile Crescent about 12,500 years ago, when nomadic hunter-gatherers settled down to become domestic agriculturists and herders. It was a watershed moment, the time when history began. And a bad move, according to anthropologist-author Jared Diamond in *Guns, Germs and Steel*. "The worst mistake in the history of the human race," he called it in a 1987 essay. "With agriculture came the gross social and sexual inequality, the disease and despotism, that curse our existence."

Building on the work of other "revisionists" (whom, uncharitably, he mostly fails to credit), Diamond made his case by citing a slew of findings that challenge the "progressive" case, that is, that agriculture gave us the "Splendors of Civilization." Instead, for Diamond, it was the start of the "Highway to Hell":

- Traditionally, farmers concentrate on growing high-carbohydrate cereals: rice, wheat, and maize (corn). Even today, we—the global population—get more than half our daily calories from such starchy grains. (In affluent countries like the United States, we also eat grain-fed animals.) Hunter-gatherers, on the other hand, have a varied and much healthier diet of plants supplemented with animal protein.

- According to anthropologist Richard Borshay Lee's work with the !Kung San (some of the world's last hunter-gatherers), these Kalahari bushmen typically spend no more than 20 hours per week obtaining food. For Hadza nomads of Tanzania, it's 14 hours or less. Tell that to your 80-hour-a-week stockbroker stuck in a traffic jam trying to get to work!
- Ancient skeletons from Greece and Turkey tell us that the agricultural settlers were weaker and shorter (height being a proxy for health) than their hunter-gatherer forebears. Men's average stature fell from 5 feet 10 inches to 5 feet 6 inches and women's from 5 feet 5 inches to 5 feet 1 inch after the revolution.
- Diamond cites studies of Native American skeletons from the Dickson burial mounds in Illinois indicating a health change for the worse when hunter-gatherers settled down some 900 years ago. Life expectancy fell from 26 to 19 years, while "the farmers had a nearly 50 per cent increase in enamel defects indicative of malnutrition, a fourfold increase in iron-deficiency anemia, a threefold rise in bone lesions reflecting infectious disease in general." (Skeletons can tell us a lot.)
- Parasites and epidemics (bubonic plague, tuberculosis, and so on) can't take hold in scattered, nomadic hunter-gatherer groups. That requires dense populations of people—towns and cities—which develop from sedentary farming societies.

Apart from all the nutritional and health changes wrought by the Neolithic Revolution, the real problem, according to Diamond, was the rise of hierarchy, haves versus have-nots: "Hunter-gatherers have little or no stored food . . . like an orchard or a herd of cows. They live off the wild plants and animals they obtain each day. Therefore, there can be no kings, no class of social parasites who grow fat on food seized from others." Veteran anthropologist Marshall Sahlins puts the issue into a broader context, arguing that hunter-gatherers took what he calls "the Zen road to affluence": they were affluent not because of how much they had but because of how little they needed.

POLYGAMY VERSUS MONOGAMY

Since Diamond popularized this "revisionist" thesis, sociologists and anthropologists have seized on it to explain everything from the subjugation of women (not generally seen in present-day hunter-gatherer groups), warfare, and—according to psychologist Christopher Ryan and psychiatrist Cacilda Jethá in their 2010 book *Sex at Dawn*—monogamy. They argue that most hunter-gatherer societies were and are polygamous, living without nuclear families and unconcerned about paternity, thus fostering intertribal cooperation and peaceful relations: you don't kill the folks who may be your immediate kin.

In an interview, Ryan put it this way: "The advent of agriculture . . . introduced the notion of property into sexuality. Property wasn't a very important consideration when people were living in small, foraging groups, where most things were shared, including food, child care, shelter, and defense. It makes perfect sense that sexuality would also be shared—why wouldn't it be when paternity wasn't an issue?" (Bonobos, our closest relatives, go at it—faithlessly—night and day while living in egalitarian, peaceful groups.)

So Diamond's theory is that the Neolithic Revolution initiated the notion of property rights, along with concern for paternal certainty and hence monogamy—especially for women. Overall, females probably came out worse than males as a result of the revolution, a point made as long ago as 1884 by Karl Marx's collaborator Friedrich Engels, who claimed that agriculture was the source of sexual inequality. (Engels was hearkening back to a mythical time when our ancestors lived in a communist utopia.) And since lactation inhibits conception, women got pregnant more frequently—with the associated health hazards of childbirth—once goat's milk and puréed cereal were available as replacements for breast milk.

With the high academic stakes involved, it's no wonder that Diamond's "revisionist" case has been challenged by "progressives," who point to all the benefits brought us by agriculture: civilization, population growth, science, and culture! (Culture, agriculture, and cultivation are cognates.) Cities, government, metallurgy, books, the Eiffel Tower, and Tchaikovsky's Symphony No. 6 didn't just happen by chance. You don't go to the moon if you're living in a small, tight-knit group, spending your days chasing animals, living in caves, and worrying about the local waterhole drying up.

And what about those peaceful relations supposedly enjoyed by hunter-gatherers? Linguist Steven Pinker, in *Better Angels of Our Nature,* argues—controversially—that we're now living in the most peaceable period in human history. Those guys back then were constantly battling, he claims, citing as evidence the frequency of weapon marks on ancient human bones.

For another even more controversial take on the "grand narrative" approach to human history, check out David Graeber and David Wengrow's *The Dawn of Everything*, which argues that it's not so cut-and-dried. In their copiously researched view, societies have always (pre- and post-agricultural revolution) experienced tension between egalitarian and authoritarian ways of group living.

Feisty lot, those anthropologists.

After all that, you may ask (paraphrasing Monty Python's *Life of Brian*), "What has the Neolithic Revolution ever done for us?" Sure, it's given us long lives, soy milk lattes, and Viagra. But it's also given us nuclear weapons, televangelism, and the Kardashians. Was Jared Diamond that far out of line in calling it "a catastrophe from which we have never recovered"?

8

ÖTZI THE ICEMAN

He didn't die of natural causes

Peering through a small, double-glazed viewing window at the short, brown, desiccated body of a man who died 5,300 years ago is an uncanny experience. Should he even be on display, for the likes of me and several hundred thousand visitors per year, in his bespoke museum in Bolzano, Italy? Since being discovered high in the Italian Alps in 1991, he's been examined more than any other human being, so a few noninvasive gazes won't make much difference. But still . . .

Ötzi (pronounced Ert-zi), the name given him by an Austrian journalist (he was found in Italy's Ötztal Alps), was discovered half in, half out of melting ice at 10,500 feet close to the border with Austria. So close that ownership of the body was contested for years by the two countries—Italy won.

As mummies go, Ötzi is unique, a wonderfully preserved "wet mummy," unlike, say, dry Egyptian mummies, whose organs were ritually removed as part of the process. Soon after his death, snow, then glacial ice, entombed Ötzi, where, happily for us, the cold kept his body from decomposing.

In life, he was a Caucasian male, about 110 pounds and 5 feet 2 inches, age 45 years at death. (His weight and height were average for his time.) He's now shrunk a little and weighs just 35 pounds. He was also a murder victim, a fact not appreciated until an X-ray taken 10 years after his discovery revealed a flint arrowhead lodged beneath his right shoulder. Ötzi probably bled to death soon after being shot from behind.

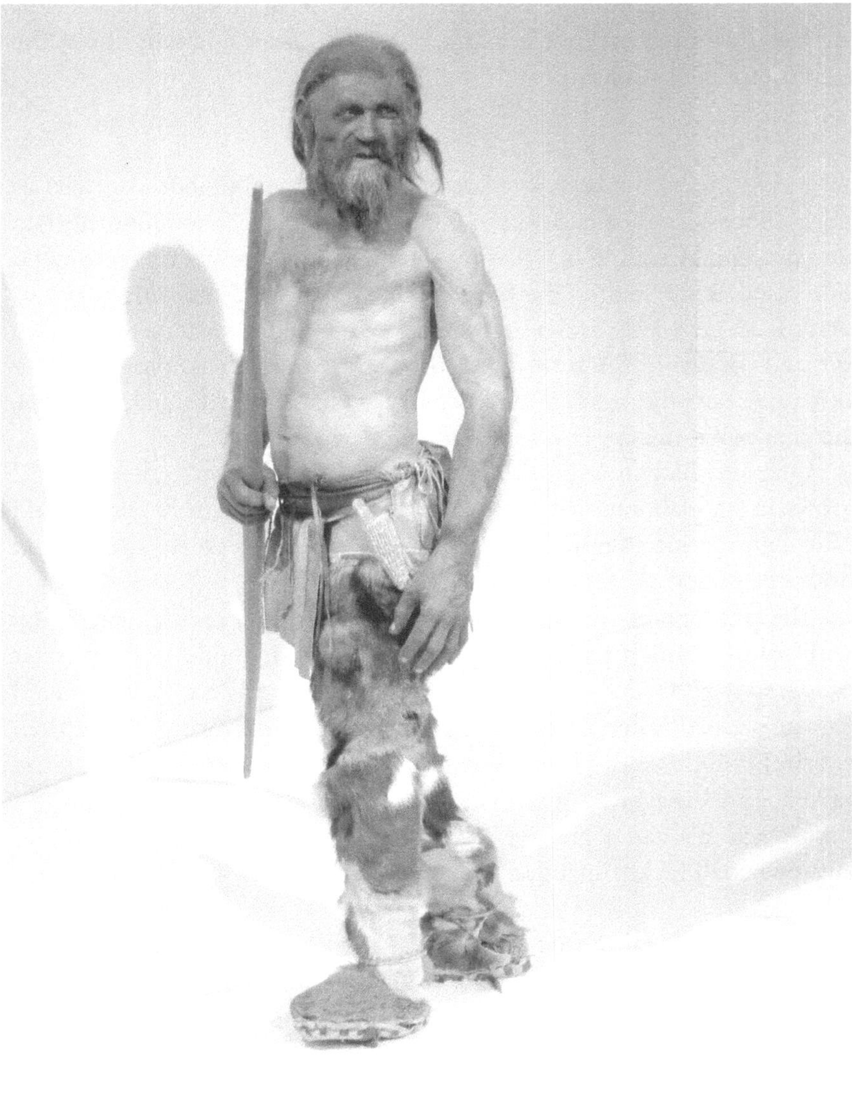

Reconstruction of Ötzi by paleo-artists Adrie and Alfons Kennis. AUTHOR

Since the contents of his stomach showed that he'd stopped to enjoy a hearty meal just a few hours before the ambush, he probably didn't realize he was in danger. And from a two-day-old deep cut on his hand, it seems he'd been in a recent fight, leading to a host of speculations about the circumstances of his murder.

ÖTZI'S KIT

As interesting as the body is to researchers, his equipment also tells us much about him. For example, a copper-headed axe was found next to him, presumably ruling out theft as a motive. Copper was the first metal to be used at the end of the Stone Age, and the 99.7 percent purity of Ötzi's four-inch-long axe-head lashed to a yew handle attests to the skill of metalworkers of the time. After the coppersmiths heated ore-bearing rock to release the copper, they would have cast, cold-forged, polished, and sharpened the axe in a series of elaborate processes.

Ötzi also had an unfinished six-foot yew longbow, together with 12 arrows in an elaborate leather quiver; a short dagger on his waist; and a fire-lighting kit comprised of a flint and pyrite (to create sparks) and fungus as tinder.

The photograph shows a reconstruction of Ötzi in life, created by twin-brother Dutch paleo-artists who based their model on computer tomography and CAT scans. Although he looks pretty hardy, he suffered from an arthritic knee, bad teeth, Lyme disease, whipworm (an intestinal parasite), and possibly a head injury. Still, his strong leg muscles tell us he walked a lot—maybe he was a goat herder or a shepherd?

Or was his axe a prestige item that only a tribal headman would have owned? His kit included some medicinal herbs, so perhaps he was a shaman.

It's weird to realize that when the Great Pyramid of Giza was being built, Ötzi had already been entombed in ice for 700 years. And he'll be around for many years to come since he's kept frozen at −6°C under controlled conditions to prevent decomposition. He probably won't mind a visit if you find yourself in the South Tyrol and may even appreciate all the attention. Personally, I'd be flattered if people paid to see my naked body 5,000 years in the future.

Anytime, really.

9

THE BRONZE AGE

Copper good, bronze better. Much better.

We humans love dividing the world up into threes: animal, vegetable, mineral; mind, body, spirit; past, present, future; starter, entrée, dessert. And so it was 200 years ago, when Christian Thomsen, curator of what was to become the National Museum of Denmark, gave the world the Stone Age, Bronze Age, and Iron Age. Within a couple of decades and despite its many flaws, European archaeologists adopted Thomsen's tripartite division of deep history to the extent that it persists even today—you probably learned it at school. It's still handy, if somewhat inaccurate, to talk, for example, of the "Late Stone Age" and the "Bronze Age collapse."

Inasmuch as the system is useful, it's the Bronze Age that warrants our attention, for it was the serendipitous discovery of bronze—at least in Europe and the Middle East—that changed everything. (In some other parts of the world, like sub-Saharan Africa, they essentially skipped bronze and went straight from stone to iron.) For some 3 million years prior, early man had chipped and flaked stone cobbles and flint shards to make crude hammers and cutting edges, with little sign of improvement from one generation to the next. But once our ancestors figured out bronze, exponential progress followed.

Bronze wasn't the first metal to be utilized for tools by our forefathers since copper has that distinction. (Previously, native gold and meteoric iron had been used only for jewelry.) For example, Ötzi the Iceman's axe-head is 99.7 percent pure copper. Copper, usually in the form of an ore but sometimes as a pure metal, is quite common, being found in Earth's crust at a concentration of about 70 parts per million. Compare that to ubiquitous iron and rare tin at concentrations of 50,000 and 2 parts per million, respectively.

Ötzi died around 5,300 years ago, while the first evidence for copper smelting—heating the ore to extract metal—dates to about 7,000 years ago. This ushered in the Chalcolithic (or Copper) Age, which didn't make Thomsen's classification, and that period now is usually considered to belong to the Neolithic, or New Stone Age. The problem with copper is that it's soft. Ötzi couldn't have downed a small tree without repeatedly sharpening his axe.

Bronze is what you get when you combine copper with either arsenic or tin in a roughly 10-to-1 ratio. It's that addition that makes all the difference because bronze is much harder and more durable than copper. For 2,000 years, from about 3300 to 1300 BC, bronze was the magic metal in Europe and the Middle East. It was used for practically everything: weapons, armor, agricultural tools, domestic items for cooking and sewing, mirrors, coinage, and sculpture.

LADY OF KALYMNOS

The gorgeous *Lady of Kalymnos* is an intriguing example of the latter. In 1994, a Greek fisherman found the larger-than-life bronze figure between the Dodecanese islands of Kalymnos and Pserimos (north of Rhodes) in 300 feet of water, where she'd probably lain for 2,000 years. She had been cast in the Hellenistic period, around the third century BC, and was probably en route to Rome when the ship in which she was being transported foundered. Notice her excellent state of preservation, thanks to bronze's property of resisting corrosion, even in seawater. She's now in the Archaeological Museum of Kalymnos in Pothia, Greece, where I took her photo.

Absent knowledge of metallurgy, how could the first bronze workers—probably Sumerians living in Mesopotamia around 3500 BC—have come up with the magic formula of adding arsenic or tin to copper? We'll

Lady of Kalymnos. The bronze statue, now in the Archaeological Museum of Kalymnos in Pothia, Greece, was found in 1994 on the seabed by a local fisherman. AUTHOR

never know for sure, but a reasonable speculation is that copper and tin ores found their way into a single firepit—perhaps they were being used as a fire ring. As the heat smelted the ores, the two molten metals combined to form bronze. Since copper melts at 1,085°C/1,984°F (and tin much lower, at 232°C/449°F), a firepit could have done the job. However it happened, we can assume that serendipity rather than foreknowledge played the major role in creating tin bronze.

Arsenic bronze is another useful metal, but early bronze makers must soon have realized that arsenic fumes were making them seriously ill, sending them to tin as the additive to turn copper into bronze. The problem is that tin is a relatively rare element in Earth's crust, and the usual ore—tin oxide, or cassiterite—is found quite erratically around the planet.

The closest sources for the ancient Mediterranean and Mesopotamian worlds were Devon and Cornwall in southwest England, Brittany in France, and the Erzebirge region on the border of Germany and the Czech Republic. By testing for isotopes, researchers can pinpoint the source of ancient bronze objects discovered in the course of archaeological excavations. It turns out that most of the tin used in Europe during the Bronze Age came from Cornish mines, meaning that a long-distance trade in tin must have flourished during the two-millennia period from 3300 to 1300 BC.

BRONZE AGE COLLAPSE

Archaeologists speculate that bronze may have been the catalyst leading to the growth of cities, the development of government, massive building programs (like the Egyptian pyramids), and—perhaps most importantly—the invention of proto-writing so that the Bronze Age collapse, usually dated to between 1200 and 1150 BC, sent the Mediterranean and Near East worlds into a tailspin that lasted several hundred years.

(The collapse occurred over a decade or so, and anthropologist Eric Cline's book title *1177 BC*, which implies it all happened in the span of 12 months, was, he admits, a marketing gimmick that his publisher came up with!)

Of the many theories proposed as the reason for the collapse, one strong competitor is the interruption of the tin trade routes, caused perhaps by a long-term drought or by relatively minor political upheavals. No tin, no bronze. And without that bronze underpinning, the entire machinery of the state was vulnerable to failure.

Following the collapse, states slowly regrouped, concurrent with the start of the Iron Age, dated to between 1200 and 600 BC depending on the region. Iron is trickier to work than bronze and has a higher melting temperature (1,538°C/2,800°F), but once efficient smelting techniques had been developed, iron became the go-to metal for most uses other than statues. For those, bronze remained the perfect material.

It still is.

10

THE LONG MAYAN DROUGHT

The end of Tikal, Palenque, Copan . . .

Great empires are traditionally founded on the banks of great rivers such as the Nile, the Yangtze, the Indus, and the Tigris-Euphrates complex because people need food and crops need water. This is why the Yucatan Peninsula of Mexico, together with the land farther south into Belize, Guatemala, and Honduras, is one of the least likely places on Earth in which to grow a civilization. Due to its porous limestone bedrock, the Yucatan lacks surface water in the form of substantial rivers and lakes. During the Mayan era, from around 3,000 years ago, agriculture was almost entirely dependent on seasonal rainfall, with limited backup from *cenotes* (natural water-filled sinkholes), *aguados* (small surface reservoirs), and *chultuns* (underground cisterns). Yet during what archaeologists have dubbed the "Classic Stage," between AD 200 and 910, the Mayan population flourished, tripling in numbers to a peak of perhaps 20 million.

Then, in the 30 or 40 years after 909 (the date of the last known "Long Count" inscription, by which they kept track of the age of their civilization), much of the Mayan society collapsed. Archaeological evidence shows that most of the major urban centers, especially in the southern lowlands, were abandoned, including the great cities of Tikal, Calakmul, El Mirador, Palenque, and Copan. What caused such a rapid and near-complete collapse? Archaeologists haven't been shy about offering explanations: by one count, no fewer than 88 different scenarios have been suggested, including tribal warfare, epidemics, foreign invasion (perhaps by Toltec or Teotihuacan people), revolution, disruption of trade routes, and royal in-fighting.

One hypothesis that has been steadily gaining traction is prolonged drought. By sampling lake-bed sediments and measuring stalagmites (whose growth varies according to moisture levels), paleoclimatologists estimate that starting around AD 800 and lasting for 200 years, the Yucatan region suffered its worst drought in 7,000 years. Without rainfall, crops failed, leading to the demise of the empire's many city-states. If this theory is correct, the wonder is that they hung on as long as they did.

A study by researcher Ben Cook from NASA's Goddard Institute for Space Studies proposes that the Mayans may have unwittingly helped to bring on their own demise. In order to grow sufficient corn and manioc in the area's thin topsoil, they probably clear-cut some 100,000 square miles of forest, equivalent to the area of Oregon. Crops have a higher albedo (reflec-

At 150 feet high, "Structure 2" is the largest building at Calakmul in the Petén Basin of the Yucatan Peninsula. The city was home to an estimated 50,000 people before being abandoned early in the tenth century. AUTHOR

Author pretending he can decipher Mayan hieroglyphics at Calakmul. AUTHOR

tivity) than forests, resulting in less sunlight being absorbed. This translates into less energy being available for convection and hence for rainfall. While not claiming that clear-cutting actually *caused* the drought—and hence the decline of the Maya—Cook says that his team's analysis shows that "deforestation can bias the climate toward drought and that about half the dryness in the precolonial period was the result of deforestation."

The *indigenas* never recovered between the tenth-century collapse and the Spanish conquest six centuries later, giving the forests a chance to regrow. Today, the vast bulk of their former territory is covered in trees. If not for their ruins, it's as if the Classical Mayan Empire never existed.

ENGINEERING

Beyond Lego® blocks

11

PYRAMID SCHEME

Does the Great Pyramid incorporate pi?

Of the 90-odd pyramids in Egypt, not to mention hundreds more in South and Central America, Cyprus, Sudan, Thailand, and Myanmar, there's only one *Great Pyramid*. It's simply the largest stone structure ever erected and the oldest of the Seven Wonders of the World. Completed around 2550 BC, it was intended to be the tomb of Pharaoh Khufu (Greek Cheops), although his body may never have been placed there. The pyramid, originally 481 feet high, consists of more than 2 million blocks of limestone and weighs more than 6 million tons. It's not, however, the largest pyramid of all if we include those not made of stone, as we'll see.

DESIGN OF THE GREAT PYRAMID

One of many (many!) controversies surrounding the Great Pyramid is its design, in particular the steepness of its faces. Did Khufu's nephew Hemiunu, whom archaeologists think was the architect, make it easy for the builders by giving them an easy ratio to follow? Or did he cunningly incorporate a mathematical constant into the design? Let's look at three possibilities:

- Golden Ratio: The Greek "father of history" Herodotus, touring Egypt 2,000 years after Khufu's reign, suggested that the area of each of the pyramid's four faces equals the square of its height. If that was a deliberate choice, then a so-called magic number, the "golden ratio" phi (ϕ, approximately 1.618), might have been in the architect's mind. Phi crops up in geometry (pentagons), nature (Fibonacci series), architecture, and art. A "phi pyramid" built according to Herodotus' suggestion would have a slope angle of approximately *51.827 degrees.* (For those mathematically inclined, the ratio, base to slope length to height, is $2:\phi:\sqrt{\phi}$.)

- Pi (π): Or maybe Hemiunu designed it as a "pi" pyramid. In such a structure, dividing the area of the pyramid's base by its height gives us pi (a circle's circumference divided by its diameter, about 3.142), resulting in a side slope of approximately *51.854 degrees.* (The closeness to the golden ratio slope results from pi being approximately $4/\sqrt{\phi}$.)
- Easy Ratio, 11/14: Yet another possibility—and the simplest for the builders—is to make the side slope 11/14, that is, 11 horizontally to 14 vertically, giving us approximately *51.859 degrees.*

Do you see the problem? The three angles are so close that it's impossible to decide which one is correct. With the original marble sheathing now missing (stripped in the Middle Ages to build Cairo's mosques and palaces) and the capstone, or pyramidion, gone, surveyors can't measure the slope with any confidence. All we can really say is that it's between 51.5 and 52 degrees, so the architect *probably* designed the pyramid

Pyramid of Khufu and the Sphinx. AUTHOR

deliberately, according to one of the above ratios. How can we be so sure? Because those ancient builders knew about precision! For example, Khufu's pyramid is aligned to within one-fifteenth of a degree to true north, astonishing accuracy considering what they had to work with back then, so we can say with confidence that the slope of the faces wasn't left to chance. Which one did they choose? We'll never know.

We can be certain, however, that the unfinished pyramid on the back of a dollar bill—part of the Great Seal of the United States—bears no relationship to phi, pi, or 11/14. The slope, despite its Masonic origins, is an unstable 70 degrees.

For an in-depth discussion, check out https://sites.math.washington.edu/~greenber/pyressay.html.

THE CHOLULA PYRAMID

I said that the Great Pyramid is the world's largest stone structure. However, the world's largest artificial structure by volume (according to the *Guinness Book of Records*) is to be found not in Egypt but rather on the other side of the world, near Pueblo, Mexico. Despite its better-known cousins, the pyramids of the Sun and the Moon at Teotihuácan outside Mexico City, the Cholula pyramid is the record holder. With a base of about 1,500 by 1,500 feet and a height of 220 feet, it has a total volume of nearly 6 million cubic yards, about one-third as much again as Khufu's pyramid.

The story goes that, after the Spanish conquered Puebla in 1520, massacring thousands of indigenous people in the process, they constructed a church on top of what they thought was a hill. Only later did they realize they were building on a "pagan" site previously dedicated to the Aztec feathered-serpent god Quetzalcoatl and that the hill was in fact a huge artificial mound. Later still—starting in the 1930s—archaeologists dug five miles of tunnels through the hill to uncover its construction. They found that it had been built in about six stages, starting around 200 BC, each stage enlarging on the previous one.

Having (illegally) climbed Khufu's pyramid, I can report that the steep path to the top of the Cholula pyramid makes for a *much* easier ascent and, critically, descent.

The Cholula pyramid with a sixteenth-century Spanish church on top. AUTHOR

Model in the adjacent museum shows what the archaeologists' tunnels revealed: several stages of construction, each one built over the previous pyramid. AUTHOR

12

MIND THE GAP!

A brief history of the arch

For this retired civil engineer, one of the mysteries of ancient civilizations is why it took so long for engineers of old to figure out the "true arch." All buildings need openings for doorways and windows, and the earliest and simplest way to span an opening is with a lintel, a flat piece of stone or wood or, today, metal. For ancient people, wood often wasn't available, or, if it was, it rotted. Steel (an alloy of iron and carbon) hadn't been invented, which left stone. Stone is usually really strong in compression but weak in tension: if you stretch it, as happens with the part of the stone on the underside of a lintel, it's vulnerable to cracking, thus limiting the width of the opening.

The ancient Egyptians had another trick up their sleeves: *corbels*. The pyramids are essentially huge tombs, so they needed to create rooms deep inside the pyramids large enough to accommodate a stone sarcophagus (coffin). Take Khufu's pyramid. Right in the center of the huge structure is the King's Chamber, a hollow space about 36 feet long by 18 feet wide. (Curiously, archaeologists believe that the huge granite sarcophagus found there was never used.) Above the chamber are hundreds of tons of limestone blocks bearing down, far too much weight for a simple stone lintel to support across the width of the chamber. So the Egyptians used a corbel system, in which successive layers of stone—in this case, granite—are stepped closer and closer together forming an inverted-V

Left: Lintel at Machu Picchu, Peru. Middle: Corbel at Uxmal, Yucatan, Mexico. Right: Roman arch at Tlos, Mugla, Turkey. ALL AUTHOR

ceiling. Much later, the Incas, Mayans, and Aztecs used the same device whenever they wanted to span more than a few feet.

Although the odd attempt to improve on these methods had been made earlier, it wasn't until the Roman Empire that we see the full flowering of the successor to the lintel and corbel: the arch. Here, all the individual stones, or *voisseurs*, are compressed, taking advantage of the ability of stone and brick to resist compression. Unlike a corbel arch, where each side is freestanding, a true arch would collapse if you removed part of it.

To construct an arch, the Romans first built a wooden "falsework" in the shape of the arch's underside on which each individual *voisseur* was positioned until they reached the "keystone" at the top. With that last piece firmly in place, the falsework was removed, and the arch, with all its elements in compression, stayed up on its own. Note that a true arch pushes outward as well as down, so the ends, the abutments, need to be sufficiently stable to support both the vertical weight and the outward force. As I say, it seems so obvious. I wonder why earlier civilizations, such as the Mesopotamians, didn't hit on it.

Although the Romans focused on the round arch, many variations are possible, all of which avoid putting stone or brick into tension. For example, the designers of the thirteenth- and fourteenth-century cathedrals of northern Europe found the pointed "Gothic" arch more to their taste. And although it's not built of stone or brick, the stainless-steel Gateway Arch in St. Louis, with its "inverted weighted catenary" profile, takes arch building to new heights (as it were) of efficiency, not to mention grace.

13

THE ANTIKYTHERA MECHANISM

A 2,000-year-old computer

For 2,000 years, it lay 150 feet below the surface of the Mediterranean Sea. Since being found accidentally by sponge divers in 1901, it's been the source of speculation and wonder to the present day. It's so unlikely that the first time I heard about it, I was convinced the whole thing was a hoax.

In fact, the Antikythera mechanism, about the size of a thick phone book, is arguably the single most impressive artifact that we have from the ancient world and by far the most elaborate artifice that we know of from any time prior to AD 1300. The divers recovered it, along with a rich trove of bronzes, pottery, and glassware, from a Roman merchant ship that had been wrecked in about 60 BC off the rocky coast of Antikythera, Greece, a small island lying between the Greek Peloponnese and Crete.

The mechanism consists of bronze gears and bearings artfully sandwiched between two inscribed brass plates, around which pointers rotated. The user turned a crank on the side to adjust the mechanism. The best analogy I can think of is a complicated eighteenth-century clock.

It's not a clock, though, but rather a mechanical calendar, that is, a computer, built between 150 and 100 BC. Now containing at least 30 precisely cut toothed gears (up to 40 more are lost), it was designed to show the positions of the sun, moon, and planets for any date as well as phases of the moon and predictions for solar and lunar eclipses.

Antikythera mechanism fragment A. The entire mechanism consists of a complex system of 30 wheels and plates with inscriptions relating to signs of the zodiac, months, eclipses, and pan-Hellenic games.

The technology of the mechanism is astonishingly sophisticated, best seen in how its designer (Archimedes perhaps?) incorporated the moon's motion. Although the ancient Greeks didn't know that the moon's orbit is an ellipse—that had to wait for Johannes Kepler nearly two millennia in the future—they did know that the moon appears to speed up and slow down in a predictable rhythm relative to the background stars. This rhythm was built into the mechanism using an ingenious pin-and-slot epicycle gear system that accurately mimicked what we now know to be elliptical lunar motion.

Subsequent to the shipwreck, the mechanism's finest hour came in 2005, when a team from Cardiff University in Wales used high-resolution computed tomography to scan it in an eight-ton machine built specifically for the job by X-Tek Systems, part of Nikon Metrology.

The mechanism, now in 82 fragments, is housed in the National Archaeological Museum of Athens. Because of its fragility, X-Tek had to transport their cumbersome device to the museum rather than the other way around. The results of that scan have revealed many more details previously unavailable to researchers, including about 2,000 Greek letters giving directions for its use. If you've ever lost a user's manual, you'll appreciate the wisdom of putting the instructions right on the gadget itself.

For dynamic simulations of how it worked, check out http://www.etl.uom.gr/mr/index.php?mypage=antikythera. More information can be found at the website of the Antikythera Mechanism Research Project: http://www.antikythera-mechanism.gr.

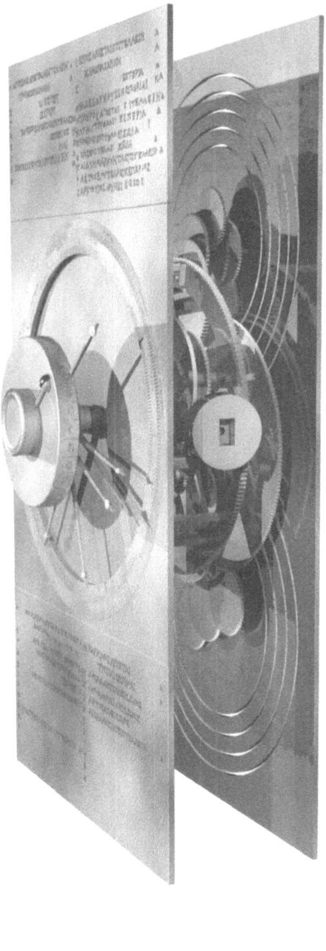

How it might have originally looked. On the left are dials for the positions of the sun, moon, and planets. The rectangular slot was for a crank that the user turned to enter the date.

14

PERRY COLLINS'S OVERLAND TELEGRAPH

A world-class miss

When I consider monumental works of humankind, I think of the Great Pyramid, the Eiffel Tower, the Statue of Liberty, the Saturn V rocket, the Panama Canal, Boulder Dam—and Perry Collins' Overland Telegraph. This last one isn't usually included in any "greatest achievements" list for good reason: it was never finished—not because it wouldn't have succeeded but because it was preempted by another wild scheme.

In the mid-1800s, with communication between Europe and America limited by the speed of sailing ships, several entrepreneurs dreamed of a telegraph line linking the Old World with the New. After several attempts to lay an undersea cable across the Atlantic had failed, Perry McDonough Collins, an American businessman with connections in both Washington and St. Petersburg, conceived of an *overland* telegraph between San Francisco and Moscow via British Columbia and Siberia.

His plan was to extend the new east-to-west transcontinental telegraph line north into the interior of British Columbia, through Russian America (future Alaska), and thence undersea across the 52-mile-wide Bering Strait to the mouth of the Amoor River. From there, it would run 5,000 miles overland across Siberia to Moscow, which was already connected by telegraph to St. Petersburg and the rest of Europe.

After years of negotiations, Collins' crews started construction in 1865. By the summer of the following year, the Canadian and Russian

Painting by John Clayton White (1835–1907) of the construction of the Collins Overland Telegraph. PUBLIC DOMAIN

sections were well on their way, and it would probably have been operational by 1867. But then fate intervened in the form of another bold entrepreneur: Cyrus Field.

Field had failed in his first attempts to lay an undersea cable across the Atlantic in 1857 and again in 1858. Finally, he leased the *Great Eastern*, a ship five times larger than any other afloat at its launching in 1858. On its third attempt, the *Great Eastern* succeeded in connecting America to Europe by copper cable, and in July 1866, the first message was transmitted under the Atlantic from Cape Clear, Ireland, to Heart's Content, Newfoundland. Eight months later, with the transatlantic cable still working, Collins stopped work on the now obsolete overland line.

One lasting consequence of Collins' aborted plan was to put Russian America onto the map and into the consciousness of the government of the United States. Without Collins' vision, the U.S. flag might have 49 stars and Alaska might still be Russian.

15

THE MOST EXPENSIVE LIGHTHOUSE

California's St. George Reef

In 1792, Captain George Vancouver, sailing north toward Alaska, gave the name "Dragon Rocks" to a treacherous reef six miles west of the nearest land that he called "Point St. George" near present-day Crescent City in Northern California. By the time the steam side-wheeler *Brother Jonathan* foundered on the same reef on July 30, 1865, three months after the end of the Civil War, navigation charts were identifying the hazard by its present name, "St. George Reef." The *Brother Jonathan* disaster, in which 225 lives were lost, prompted the U.S. Lighthouse Board to request funds to build a lighthouse on the reef. It took 17 years for the initial funds to come through and another 10 before the light shone for the first time.

The direct predecessor of the St. George Reef light was the Tillamook Rock lighthouse off the northern Oregon coast, which Scottish engineer Charles Alexander Ballantyne completed in 1881. The next year, supported by a $50,000 appropriation from Congress, Ballantyne surveyed the reef and chose one-acre Northwest Seal Rock as the best construction site for what would be the most inaccessible lighthouse in the United States. The toughness that made the metamorphic rock so desirable as a foundation would also be a nightmare to drill. However, in one hectic summer season of work, drillers and dynamite blasters completed the base for the pier. By October 1883, when winter storms forced a hiatus in construction, the previously smooth rock had been sculpted into a stepped pyramid core ready to receive the first granite blocks for the lighthouse base.

HUMBOLDT COUNTY GRANITE

Ballantyne originally thought that granite would have to be shipped all the way from San Francisco Bay 300 miles to the south. However, in December 1883, he learned of a much closer source of first-quality granite on Bella Vista Hill, on the north bank of the Mad River, in Humboldt County, California. The location could hardly have been more convenient, since lumberman John Vance had recently completed his Mad River Railroad, which ran right past Bella Vista Hill.

Ballantyne contracted an ex-captain of the Army Corps of Engineers, A. H. Payson, to quarry huge chunks of granite and skid them down the bottom of the hill, from where they were loaded onto railroad skips, thence by rail to the Mad River Slough. They were then loaded onto barges, skips and all, to be towed down Humboldt Bay to what became known as Paysonville, on Humboldt Bay's North Spit.

It's a pretty idyllic spot now, but it must have been a scene of wild activity for the three years from 1884 to 1887. Payson employed a gang of Italian stonecutters to offload the raw granite blocks and cut them down to exact size, allowing a tolerance of no more than three-sixteenths of an inch (in tough-as-nails granite!). The finished blocks weighed nearly three tons on average, while the heaviest tipped the scales at twice that.

CONSTRUCTION

Construction of the lighthouse proper began in the spring of 1884. The first batch of a total of 1,339 dressed granite blocks was winched from a ship to freshly hewn Northwest Seal Rock, where workers anchored the lowest layer of stones to bare rock with two-and-a-half-inch-diameter brass dowels. By the end of that construction season, 13 courses had been set in place with sturdy mortise-and-tenon joints.

I imagine that at that point, Ballantyne and his crews were anticipating a rapid completion to their project. Fate, however, intervened in the form of the 1884 "Grant's Panic" stock market crash. Funding dried up. Congress appropriated a measly $40,000 in 1885 and zip in 1886.

The following year presumably saw an improvement in the economy since Congress released $120,000 to continue, and from there on, work proceeded rapidly. The 50-foot-high pier was finished in 1889, with construction of the tower complete by the end of the 1891 construction season. The following year saw the arrival, from France, of nearly 700 polished glass prisms and lenses, to be assembled into the first-order (nine-foot diameter) Fresnel lens. On October 20, 1892, the lamp, 144 feet above the waterline, was lit for the first time.

THE COST

Among the superlatives associated with the St. George Reef light is "most expensive." Depending on which relative measure is used, its $752,000 price back then is equivalent to between $20 million and $100 million today. That's five times more expensive than Ballantyne's Tillamook Rock lighthouse and more than the combined cost of base construction and erection of the contemporaneous Statue of Liberty. Quarrying, dressing, and transporting the Humboldt County granite blocks accounted for about half the total cost.

Conditions for the six-man lighthouse crew were incredibly tough, and it was one of only three lights in the United States to warrant the

St. George Reef Lighthouse, six miles off Crescent City in Northern California. The 1,339 blocks of granite came from Humboldt County to the south. BRYANT ANDERSON/*DEL NORTE TRIPLICATE*, USED WITH PERMISSION

maximum "hardship pay." Few keepers signed on for a second tour of duty. Winter storms must have been especially grueling when high seas prevented supply ships from venturing out of Crescent City Harbor. During one of these storms, in 1952, crashing waves hurled rocks through the lantern room glass. Keeper Floyd Shelton is quoted in Dennis Powers' 2007 book *Sentinel of the Sea*: "There was a waterfall running down the central spiral staircase from the ocean coming in the lens room 145 feet or so above sea level."

In 1975, a large navigational buoy ("floating lighthouse") replaced the light. Subsequently the Del Norte County Historical Society removed the Fresnel lens for reassembly in their museum in Crescent City. In 1996, the St. George Reef Lighthouse Preservation Society gained jurisdiction over the site and began the arduous task of restoration, including rebuilding the lantern room. Today, visitors can occasionally tour the lighthouse after being ferried out by helicopter from the nearby Crescent City airport—a 10-minute ride that lighthouse keepers of old would have marveled over.

16

PANAMA: LINKING THE OCEANS

Saving 5,000 miles of travel

The Panama Canal is a 50-mile-long waterway connecting the Atlantic and Pacific oceans. When opened in 1914, the canal drastically changed global commerce by cutting 5,000 miles off the dangerous "Round the Horn" route. However, if Ferdinand de Lesseps, designer of the Suez Canal, had succeeded with his second great canal venture, Panama would have opened 10 years earlier under the tricolor flag of France rather than the Stars and Stripes.

The French attempt was abandoned in 1897 after 13 years, nearly $7 billion in today's dollars, and the deaths of tens of thousands of laborers (see https://pancanal.com/en/the-french-canal-construction). Why did the subsequent American attempt succeed? Three words: *design*, *machinery*, and *sanitation*:

- Design: Like Suez, the French design called for a sea-level ocean-to-ocean connection, without the complication of locks. In 1904, when Washington paid $40 million to acquire the French equipment and excavation, the U.S. engineers also envisaged a sea-level canal. However, they were soon convinced that excavating the eight-mile-long "Culebra Cut" to sea level would be a herculean task, and they redesigned the canal to incorporate locks, thereby reducing excavation to manageable volumes. Coming in from the Pacific, two sets of locks raise ships 85 vertical feet to the level of

Lake Gatun, the world's largest artificial lake in 1908, created by damming of the Chagres River. A third set of locks drops ships back to sea level at the Atlantic end of the canal.

- Machinery: Canal builders from the United States used excavation equipment that dwarfed the French machinery of 20 years earlier. The largest excavators, including giant 100-ton-capacity "Marion" (for Marion, OH) rail-mounted steam shovels, were specifically designed and built for the project.
- Sanitation: After the French failure had taken the lives of 22,000 workers (mostly victims of yellow fever and malaria), U.S. authorities took full advantage of Cuban doctor Carlos Finlay's controversial claim in the late 1800s that mosquitoes are the primary transmitters of yellow fever. Before work began, U.S. public health officials, under the leadership of U.S. Army physician William Gorgas, undertook a huge sanitation program to minimize diseases. This included installing piped water systems, fumigating buildings, installing screens and mosquito nets, draining stagnant-water breeding sites, and creating an oil film on standing water to prevent mosquito eggs from hatching. Mosquito-borne diseases, which until then had been responsible for the annual death of about 10 percent of the workforce, were virtually eliminated within two years.

The 10-year project cost nearly $10 billion in today's dollars. After 84 years of U.S. control, the canal was handed over to the Republic of Panama on December 31, 1999, and today, the future of the canal is assured, with the millionth ship passing through the original canal in 2010.

Until the opening of the parallel, wider waterway in 2016, the canal was virtually maxxed out with about 40 vessels (mostly container ships) making the 20- to 30-hour passage daily. That works out to about 300 million tons of cargo annually. The new canal alongside the original now allows passage in less than 12 hours, vastly increasing Panama's capacity.

Chances are that many of the products in your home—electrical appliances, clothing, wines, fruit, and perhaps even your car—came to you via the Panama Canal. It's hard to overestimate the influence the canal has had on world trade during the past century.

A Panamax container ship makes its way through the original Miraflores locks in 2014 before construction of the new ones had been completed. Panamax vessels were built specifically to *just* pass through the old locks with barely a foot of room on each side. A ship this size would pay about $300,000 per passage. AUTHOR

The Culebra Cut, 1896. Inadequate machinery, such as the undersized dump cars, locomotives, and excavators, contributed to the failure of the French attempt. PUBLIC DOMAIN

EVOLUTION

The single best idea ever

17

YO' MAMA'S A FISH!

Give or take 100 million generations

It's a lot of galumphing, wriggling, slithering, huffing, flopping.
—Ben Otoo, University of Chicago, on the transition from fish to amphibian

In his book *Darwin's Dangerous Idea*, now about 30 years old, the late philosopher Daniel Dennett claimed that evolution by natural selection deserves the prize for "the single best idea anyone has ever had." It's a formidable declaration, but I've never heard a worthy counterclaim.

Just to remind anyone who might be a tad rusty on their high school biology classes, what we're talking about here—evolution—is the result of random genetic mutations in the process of passing the genome (the genetic information carried in the egg or sperm) from one generation to the next. Some small fraction of these mutations are beneficial in that the members of a species that possess them are better adapted to their environment than their sibs. This gives them the edge in terms of survival and—more importantly—reproduction over those that didn't get that mutation, the sole criterion of natural selection being the passing on of genes. If the genes get passed on, the mutation was useful by definition.

We humans, together with earthworms, *E. coli*, redwoods, and fungi—everything alive—came from the same great-great- (times a billion or so) grandparent. Every member of every species (2 million classified ones plus uncountable others) is the end point of an unbroken chain of successful ancestors going back nearly 4 billion years, where "success" simply means they reproduced. We're just the vehicles for genes to unwittingly maintain themselves, hence the title of evolutionist Richard Dawkins' 1976 book *The Selfish Gene*.

TIKTAALIK

For me, what makes this spectacular process so hard to appreciate are the change points. At what point did my great-great- (times 100 million) grandmother, who looked somewhat like the Tiktaalik in the illustration, become fully amphibian, then land based, thence into a shrew-like mammal? And when did that creature eventually morph into an ape, which then, at some point, evolved into someone like Lucy (3-million-year-old Australopithecus), who became (again, at what point?) human, someone who, suitably attired and trimmed, would look like any one of us strolling down Main Street on a Sunday morning?

Trick questions. There were no "points." Pick, at random, any member of this long, unbroken chain of creatures. Now check out their parents and their immediate offspring. They all look about the same. Evolution is *gradual*, meaning it's imperceptible from one generation to the next.

Dawkins has a thought experiment on these lines, outlined in his 2011 book *The Magic of Reality* (written "for young people"—my kind of book!). He asks me, the reader, to imagine a photograph of one of my parents, say, my mother. On that photo, I place a photo of my mother's mother. On top of *that*, I place a photo of her mother—and on and on, a vast column of photos, each photo sitting between a mother and a daughter, back to nearly 400 million years ago. Allowing an average of four years between generations, that's 100 million photos, a pile as high as three Mount Everests stacked on top of one another.

Tiktaalik, our ancestor, from about 375 million years ago, is shown on the topmost photo. Tiktaalik was a fish that, over time, evolved wrists and ankles enabling it to use its fins as feet, crawling up from the water to dry land. Tiktaalik (or its close cousin) is the ancestor not only of humans but also of all non-fish vertebrates.

The painting that launched 1,000 memes. Zina Deretsky created this much-reproduced image for the National Science Foundation following the 2004 discovery of about a dozen fossils of this creature in the Arctic Nunavut Territory. Nearly 400 million years old, *Tiktaalik roseae* is a "missing link" between sarcopterygians (bony fish with lobed, paired fins) and tetrapods (four-footed land animals). Tiktaalik, meaning a "large freshwater fish that lives in the shallows," was named by local Inuits. PUBLIC DOMAIN

But there never was a point at which you could say that this generation is fish and this next generation amphibian; the changes from one generation to the next were too subtle. Similarly, from ape to human. An ape never gave birth to a human. It was all *very* gradual. But that's all it took thanks to the magic of evolution, millions of tiny changes over eons of time to get from Tiktaalik to you and me.

For more, I recommend *The Ancestor's Tale* by Richard Dawkins and Yan Wong and *Your Inner Fish* by Neil Shubin.

18

GENETIC VERSUS CULTURAL EVOLUTION

In two PowerPoint slides

Let's start with a few bullet points to sum up the major steps in the history of life on Earth à la PowerPoint:

- Genetic Evolution (first slide)
    - 4.5 billion years ago, Earth coalesced out of dust and gas surrounding a young star we'd call the sun.
    - 3.8 billion years ago, life arose in the form of simple replicators (possibly as early as 4.2 billion years ago).
    - 2.3 billion years ago, the Great Oxygenation Event began, leading to our present 21 percent oxygen atmosphere.
    - 1.5 billion years ago, single-celled organisms with nuclei (eukaryotes) were synthesized from simpler cells that lacked nuclei (prokaryotes).
    - 1 billion years ago, multicellular organisms arose.
    - 540 million years ago, early plants and animals emerged from the "Cambrian explosion."
    - 7 million years ago, our "hominid" lineage split off from that of chimpanzees.

- Cultural Evolution (second slide)
  - 200,000 years ago, modern humans appeared.
  - 100,000 years ago, language arose (a date much debated).
  - 10,000 years ago, hunting and gathering morphed into agriculture.
  - 5,000 years ago, the first writing appeared.
  - 500 years ago, the printing press was invented.
  - 40-odd years ago, the internet connected the world.

That's the quick-and-dirty story of evolution. The point is that a mere 200,000 years of *cultural evolution* makes 4 billion years of *genetic evolution* look a bit pokey. Suddenly (relatively speaking), new inventions and ideas could spread through a population virtually instantaneously. Knowledge of a poisonous fruit or improved arrowhead could be transmitted to the entire tribe in one evening! Today, the tribe is the whole world, and a breakthrough in, say, solar cell design, whisks around the internet at the speed of light. Evolutionary biologist Mark Pagel puts the emergence of cultural evolution this way: "Now ideas could arise, and they could jump from mind to mind, without genes having to change . . . cultural complexity could emerge and arise orders and orders of magnitude faster than genetic evolution."

The astonishing thing is the speed with which this happened. Two hundred thousand years represents 0.01 percent of the span of life on Earth (equivalent, for example, to the last three verses of Revelation compared to the entire Bible). Because of species' ability to not only learn from others (which dolphins and apes can do) but also pick and choose what's worth learning, human cultural evolution has made genetic evolution almost irrelevant. Once it led to creatures that could trade *ideas* in addition to genes, "regular" evolution took a back seat, while, for better or worse, we showed the rest of creation what a smart, flexible, problem-solving brain could do.

Pagel has a worrying big-picture take on the whole process. Early on in cultural evolution, while we were living in tribes of maybe 100 or so, it took one inventor to "infect" the rest of the tribe with, for example, her or his new method for skinning an antelope. That's a ratio of 1:100 "inventors" versus "copiers" per invention.

Today, we need just one innovator to come up with a great idea since other Earthlings can copy it by simply tapping into Google. So now we have a per-invention ratio of 1:8,000,000,000 inventors to copiers. Since it's so much easier to copy than to innovate, says Pagel, we're dumbing down, breeding ourselves into a species consisting of a very few brilliant inventors and the rest of us: billions of mindless copiers.

And then ChatGPT came along to complete the process.

19

THE UNITY OF LIFE

If it's alive, it's your cousin

To parody the old song, *Life is a many splendored thing*. An alien from a faraway galaxy may or may not agree with the "many splendored" part but would surely concur that terrestrial life is *a* thing, singular. Every shred of life on Earth (2 million species we know about plus maybe another 50 million we don't)—plants, animals, bacteria, fungi, and all the rest—derives from a single organism that began replicating nearly 4 billion years ago.

Our ur-ancestor is shown at the very bottom of the black line in the accompanying "phylogenetic" diagram. Since then, evolution has gifted us the wild exuberance of life that's around, above, below, and inside us. Dozens of lines of evidence lead to the conclusion that all species are interconnected:

- All organisms use the carbon-based DNA molecule, based on four nucleobases (CGAT, that is, cytosine, guanine, adenine, and thymine), to transmit genetic information from one generation to the next.
- Despite the existence of about 140 naturally occurring amino acids, all living organisms select from a set of just 23 of these.
- All these amino acids are left-handed; that is, they spiral around to the left. If you cook a batch of amino acids in the lab, you'll

get a 50-50 left-right mixture, but only left-handed ones are found in nature.

- All life on Earth is uniquely powered by charged hydrogen atoms packed into ATP (adenosine triphosphate) molecules.

If life had arisen multiple times instead of just once, we'd expect to see variance in the above, such as the occasional right-handed amino acid or another genetic transfer mechanism besides DNA (for example, RNA, which is similar to DNA except it substitutes uracil for thymine).

When we attempt to trace our family tree back to its origin (that is, LUCA, for last universal common ancestor), we find that microorganisms called "extremophiles" dominate the earliest forms of life. The "extreme" part of the term implies that they live in hostile environments—only from our point of view, of course. These little critters thrive just about

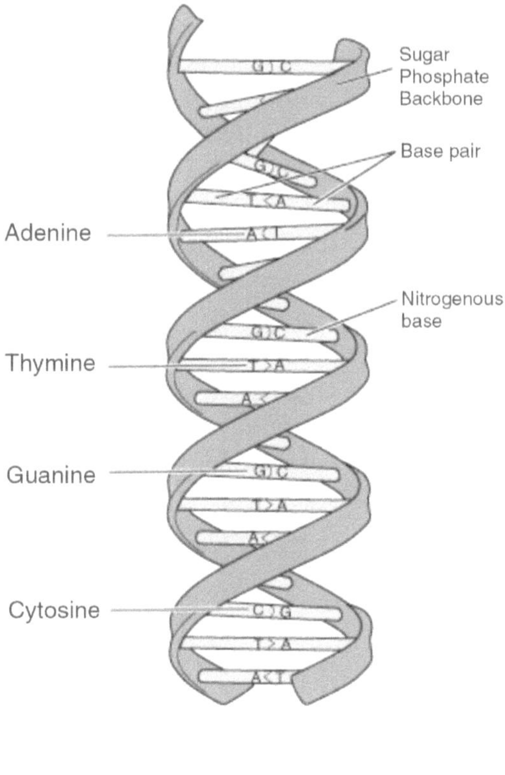

Simplified structure of DNA.
WAYBACK MACHINE, PUBLIC DOMAIN

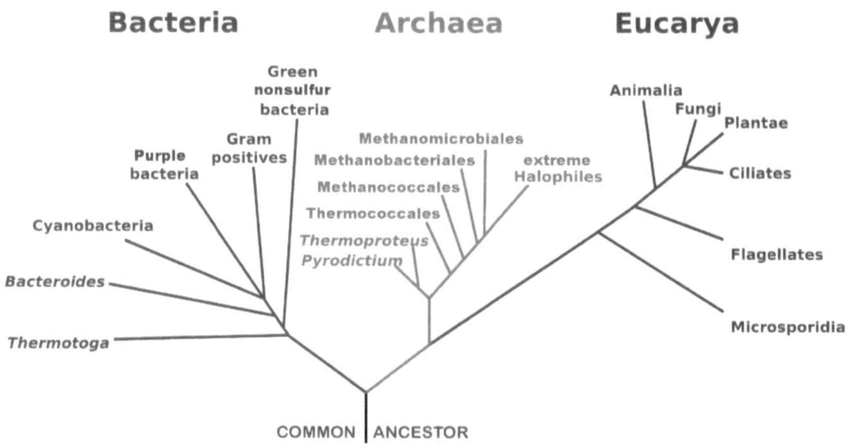

anywhere. Sites with strong acids, alkalis or radiation, boiling springs, frozen Antarctic lakes, deep underground caverns: bring it on! After all, that's where these guys came from—think of Brer Rabbit, "born and bred in a briar patch." They didn't adapt to Earth's early environment; rather, they sprang from it. Later, what we think of as "ordinary" life descended from them, as the diagram shows. (Humans come under the tag "animalia," just one twig on the profuse "eucarya" branch.)

The unity of life on Earth is what makes the search for it elsewhere so compelling. Finding life in, say, the oceans beneath the ice of Jupiter's moon Europa or in the hydrocarbon lakes of Saturn's Titan would double our sample size from one to two! And if we *do* find extraterrestrial life, what will it look like? How will it reproduce? Will it be based on carbon, like ours, or on silicon or? Are its basic molecules right- or left-handed? Are we sure we'll even recognize it as alive?

All this is moot, of course, if life in the galaxy is unique to Earth. But given how fast life arose on Earth once it had cooled off some 4 billion years ago, I'm optimistic we'll find *something* out there. The question is, what?

20

EVOLUTION ISN'T PROGRESS!

The meme that just won't quit

One of the most intriguing, and most misleading, drawings in the modern history of science.

—Science Writer Jennifer Tucker

You've seen it everywhere, in cartoons, on T-shirts, even in books on popular science. It usually starts off with a fish or a knuckle-walking chimpanzee-like creature and culminates in a sturdy guy (it's always a guy) striding manfully into the future. The message is that we—you and I—are the crown jewels of evolution, the inevitable result of a series of progressive steps from lesser creatures. Except we're not. Evolution is not progress. There's no goal, no purpose, no plan, no direction. Easy to say, hard to believe. We humans are the last to see the randomness of evolution because we think we're special. As the late anthropologist Ruth Benedict reminded us, "Fish are the last to see the water."

True, humans have qualities possessed by no other species. From our inevitable anthropocentric point of view, it may seem that skills such as technology, language, writing, and the ability to survive in any climate (even the non-climate of space) make us the top dogs here on Earth. That's the message of the "Pedigree of Man" illustration from 1879, in which we occupy the spot usually reserved for the angel on top of the Christmas tree.

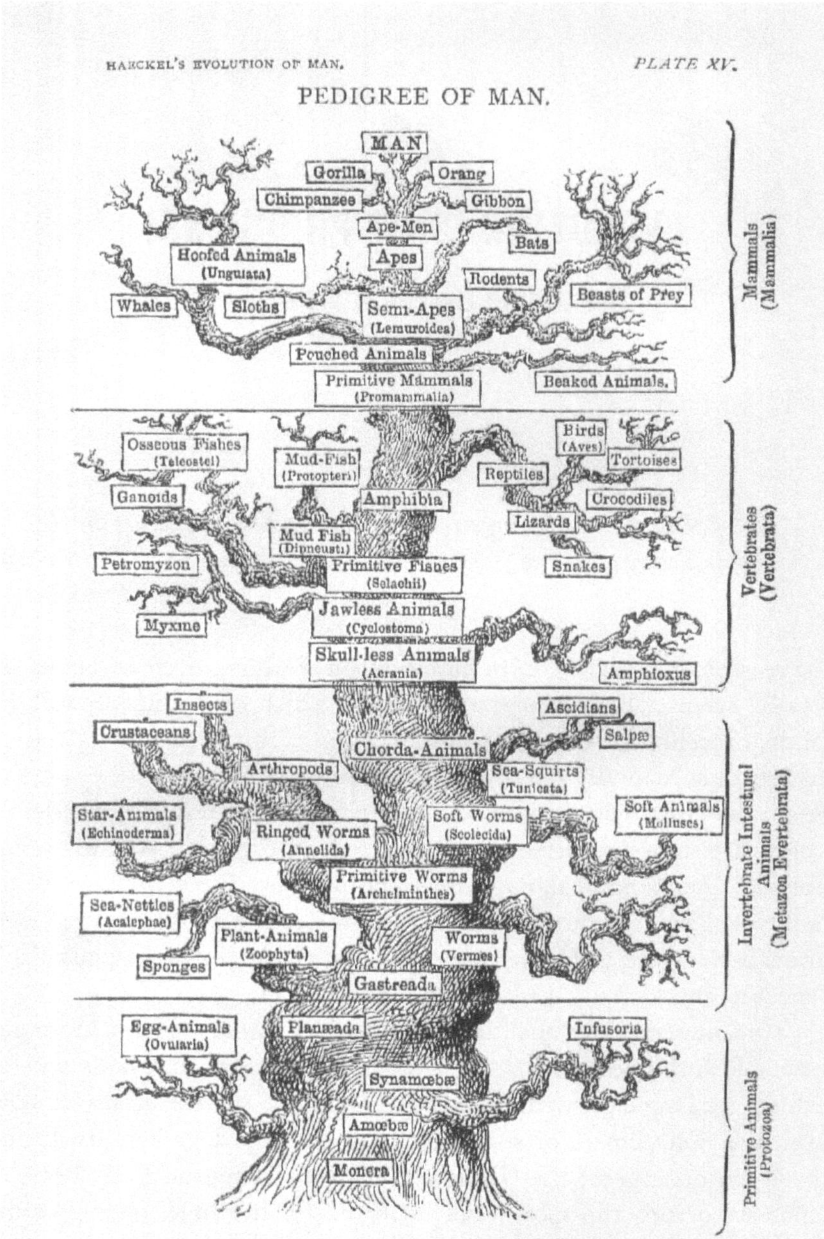

"Pedigree of Man" illustration from 1879 by German zoologist Ernst Haeckel.
PUBLIC DOMAIN

From an alien visitor's point of view, we are indeed special if the exponential growth of our species' numbers and wholesale manipulation of the global environment count for anything. But maybe ETs would deem other qualities more important if they got around to comparing species:

- Longevity: Compare our mere 200,000 years to that of ants, which have thrived unchanged for nearly 100 million years (and which don't have the self-destructive tendencies that we do).
- Ecological balance: While we ravage our planet, other species maintain a rough equilibrium with their environment.
- Coexistence: While other animals fight for resources or mates, they rarely kill their own species. We go to war.

So if evolution isn't progress, what is it? It's the process by which living organisms diversified from earlier forms and is dependent on three operations:

- Heredity: Genes are passed on to the next generation via mitosis or sexual reproduction.
- Variation: Random mutations occur when the reproductive process isn't 100 percent perfect.
- Selection: "Survival of the fittest," that is, the mostly nonrandom process by which those mutants better suited to their environment than their fellows have better reproductive success. ("Survival of the fittest," coined soon after the publication of Darwin's *On the Origin of Species*, is perhaps better phrased as "survival of the fit-enough." "Fittest" in this context simply refers to the ability to pass on genes to the next generation, so qualities such as the ability to cooperate or adapt may be more important than strength or speed.)

When challenged to summarize all the above on a T-shirt, evolutionist Richard Dawkins quipped, "Life results from the non-random survival of randomly varying replicators." Nice. Again, nothing in how evolution works implies purpose or progress. Instead, it's opportunistic, moment by moment, day by day, year by year, generation by generation. No plan, no end goal.

CETACEAN EVOLUTION

If you're unconvinced, consider the evolution of cetaceans. Some 50 million years ago, a hippopotamus-like creature (*Pakicetus* perhaps) returned to the ocean from which its ancestors had emerged a couple of hundred million years earlier. Arms and legs morphed into the fins and vestigial hind limbs in today's whales and dolphins. (The up-and-down motion of their tails belies their mammalian ancestry—fish swim side to side.) So where's the progress? Leaving the water? Returning to the water? Evolving limbs? Reabsorbing limbs?

If you think that the process by which our cetacean cousins lost their limbs is "de-evolution," you're still stuck with that "tree of life" model, in which each successive generation is an improvement on the previous one. *Pakicetus* returning to the ocean to become whales and dolphins is as much evolution as *Homo erectus* morphing into *Homo sapiens*. It's all nondirectional evolution, and it doesn't necessarily lead to greater complexity. Only when we see ourselves as special, as the pinnacle of nature's 4 billion years of experimentation, are we stuck with the fallacious notion of "progress."

PROGRESS TOWARD PERFECTION?

The seductive notion of *evolutionary progress* has persisted since Charles Darwin published *On the Origin of Species* in 1859. Even Darwin didn't totally accept evolution as a random, nondirected process, writing, "As natural selection works solely by and for the good of each being, all corporeal and mental endowments will tend to progress towards perfection." And, starting with Thomas Huxley's ape-to-human illustration four years later, we've been subjected to countless images that reinforce the notion of evolution as progress, including the image I've included here from the *Life* Nature Library series (whose editors should have known better).

So the mechanism of heredity has neither memory nor ability to see the future, which means there's no "tree of life" with humans at the top.

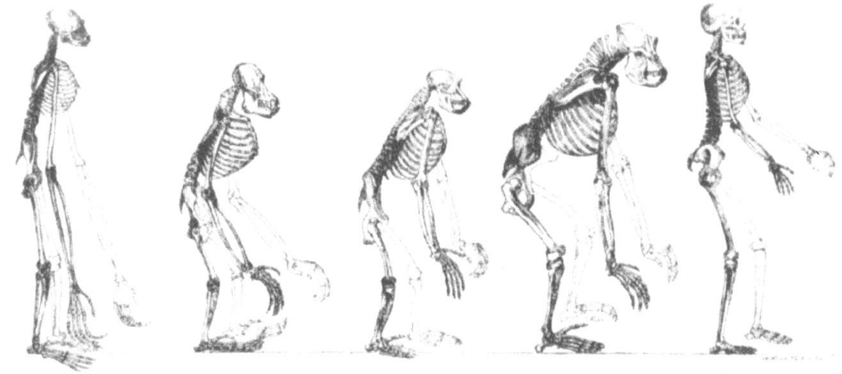

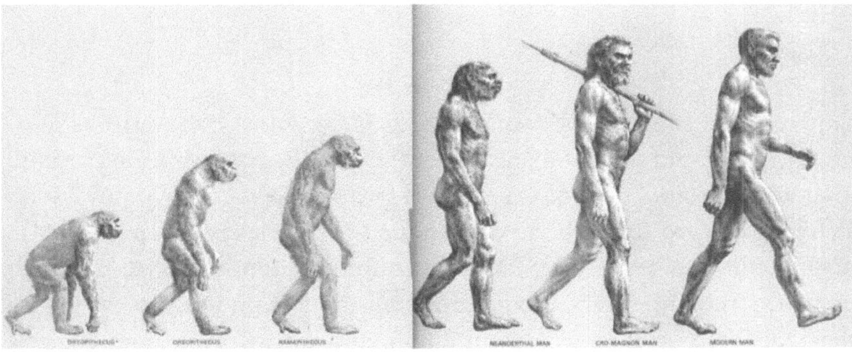

Top: When Thomas Huxley included this illustration in his *Evidence as to Man's Place in Nature* in 1863, he inadvertently gave us the template for endless "evolution = progress" memes. PUBLIC DOMAIN. Below: "The March of Progress" presents 25 million years of human evolution. This fallacious image was created for the *Life* Nature Library, published in 1965. (Fair use, low-resolution image)

A better but still imperfect metaphor is that of a bush with many living twigs and many, many more extinct twigs. Whether we're talking about original Darwinian evolution or the modern evolutionary synthesis of the 1940s that made biology a professional science, forget the T-shirt caricature. Evolution is a (drumroll) random process!

21

ARE MEN (AND WOMEN) FROM MARS?

Life began . . . where?

Suppose, just suppose, we found life on Mars: microscopic critters happily wiggling and metabolizing under the polar ice or swimming around in an underground pool of water on the warm flanks of Olympus Mons. What would we find when we examined them close-up? In particular, how would they reproduce? Would it mimic life on Earth, all of which uses the structure of DNA to get from one generation to the next? (From the DNA's point of view, of course, humans, worms, grass, and bacteria are its vehicles to survive indefinitely from one generation to the next.)

A general belief exists among "exobiologists" that any life we find elsewhere would use some other system to reproduce, similar in mechanism to DNA but with a different architecture. Maybe it would use silicon rather than carbon on which to base its cellular structure, assuming it had cells.

But suppose putative Mars life *did* have DNA similar to that on Earth. Such a finding would give a boost to the idea that, 4 billion years ago, life on Earth actually originated on Mars. A new twist to this story showed up a few years ago in science blogs. Dr. Steve Benner is a chemist and distinguished fellow at the Foundation for Applied Molecular Evolution in Gainesville, Florida. He claimed to have solved a long-standing problem with life starting here on Earth. Creation of single-stranded RNA, long presumed to be the precursor of twin-stranded DNA, almost certainly requires the presence of borate and molybdate, minerals of the elements boron and molybdenum, respectively. For a variety of reasons, the conditions on early Earth were unfavorable to the production of either mineral, unlike on early Mars, where they probably did form. So the first genetic molecule—the one from which all life on Earth descended—may have come from Mars. According to Benner, "The evidence seems to be building that we are actually all Martians, that life started on Mars and came to Earth on a rock."

EARLIEST LIFE

Let's step back and consider what the earliest life here on Earth was. Since Darwin, scientists have been struggling with that question. (Before him, there was nothing to explain—God did it.) When Darwin posited that species evolved from prior species, he implied there once was a single organism from which all species are descended. We can now say with some confidence the following about such an organism:

- It was a single cell, the simplest possible entity that can reproduce.
- It lived about 4 billion years ago.
- It used RNA (a DNA precursor) to reproduce itself.
- It made use of the essential "CHNOPS" elements we see in all life today: carbon, hydrogen, nitrogen, oxygen, phosphorus, and sulfur.

Several complicated steps were needed to get from these six constituent elements to that original living—that is, reproducing—cell. First, the elements had to combine into simple molecules such as HCN (cyanide), CH_4 (methane), and CH_2O (formaldehyde). That's easy, and so is the next step, synthesizing these rudimentary molecules into the "nucleotide bases" found in the double-helix structures of RNA and DNA: adenine, thymine, uracil, guanine, and cytosine (A, T, U, G, and C). In addition to these bases, DNA and RNA molecules have "backbones" made of phosphates and sugars. With DNA, the sugar is deoxyribose (the "D") with bases ATGC; with RNA, the sugar is ribose (the "R") with bases AUGC.

So far so good. Now comes the *really* tricky part: getting from chemistry to biology. It's not quite miraculous but is so highly improbable that creating strands of RNA out of "prebiotic" molecules in just the right formulation and geometry may have happened just once. Back to Steven Benner and his belief that planet Earth was an unlikely place for all this to have occurred. He cites three problems:

- "Naked" nucleotide bases normally decompose spontaneously, before any further reactions can occur. *Solution*: Borate (roughly, boron plus oxygen) can keep the bases stable long enough for the assembly process to take place.
- Ribose requires a catalyst to help its constituent elements self-assemble. *Solution*: An oxide of the element molybdenum can act in that helper role.
- Ribose, phosphate, and the nucleotide bases into RNA can't occur in the presence of water (the bonds aren't stable), so the assembly won't happen in, say, oceans. *Solution*: Dry land.

SO . . . MARS?

RNA probably wasn't created here because the surface of early Earth was all oceans, and boron and oxygen were in short supply. Mars is a much more likely candidate for life's cradle. Back then, Mars had far less water than early Earth, while borate minerals and oxygen were comparatively abundant, according to both evidence from our rovers and orbiters and from analyses of 30-odd meteorites researchers believe came from Mars.

Knowing that life can survive within the interiors of meteorites for eons, we can conjecture this sequence:

- RNA life began on the surface of Mars about 4 billion years ago.
- Asteroid impacts threw up showers of Martian rocks and dust, some of which escaped the low gravity of Mars, to begin long, spiraling sunward journeys.

- Millions of years later, Earth's gravity captured Martian rocks that fell to Earth as meteorites, some of which contained the seeds of life.

Life arrived, flourished, and evolved into the millions of species now living here, not to mention countless extinct ones. If this scenario is correct, when we humans eventually walk on Mars, we won't be strangers in a strange land.

We'll be home.

22

EOHIPPUS TO PERCHERON

Evolution of the horse

Soon after Charles Darwin published *On the Origin of Species* in 1859, Thomas Huxley ("Darwin's Bulldog") claimed that an unbroken 52-million-year-long sequence of fossils, cumulating in the modern horse, demonstrated unarguable proof of evolution. If only it was that easy! Since the discovery of the original succession of proto-horse fossils by American paleontologist Othneil Charles Marsh in Wyoming's Wind River basin in the 1870s, the number of "equid" fossils has multiplied exponentially. Today, the progression from the earliest proto-horse, *Hyracotherium*—popularly called "Eohippus," or "dawn horse"—to the modern horse is seen as much more complex and multi-branched than the Huxley's 1-2-3 version. The modern horse is one of the few surviving twigs on the messy "bush" of evolution.

In addition to the great difference in size and weight (Eohippus weighed about 50 pounds, while a modern Percheron draft horse typically weighs in at around 2,000 pounds), horses today walk differently than their ancestors. A modern horse walks on what was once the middle toe, while the "splint bones," metacarpal remnants of Eohippus's second and fourth toes, can still be seen as "buttons" a few inches above the fetlock.

As long as the great continent of Beringia connected Asia and the Americas, horses could have roamed freely between what we now think of as the Old and New Worlds. Based on fossil records, the genus *Equus* (which includes modern horses, zebras, and asses) originated in North

Fresh from the shower: Buster, an 18-hand (72-inch-tall), 18-year-old Percheron draft horse (AUTHOR), with Eohippus superimposed at the same scale. HEINRICH HARDER, PUBLIC DOMAIN

America about 4 million years ago and spread to Eurasia within a million or so years, with additional to-and-fro migrations.

These movements ceased when rising sea levels repeatedly fissured the two landmasses, creating the Bering Strait. The last time this happened was about 12,000 years ago. Within 1,000 or so years, horses, mammoths, ground sloths, and other large mammals went extinct in the Americas, perhaps due to a geological cooling period or to overhunting. Or it may well be that both the changing environment *and* the pressure of human hunting contributed to extinction.

The next time these shores saw horses was probably in AD 1493, transported across the Atlantic by Columbus on his second voyage to the "Indies," followed by Cortez and other conquistadores. Within a few decades, herds of feral horses were roaming the plains, and horses became an essential part of the cultural history of modern North and South America.

23

DARWIN'S BULLDOG

Thomas Huxley and the Great Debate

I was the most popular man in Oxford for full four & twenty hours.

—T. H. Huxley after his debate with Bishop Wilberforce

When, at age 25, Thomas Henry Huxley returned from a four-year stint in the Royal Navy as assistant surgeon on a Royal Navy corvette exploring the South Seas (1846–1850), he was something of a celebrity. His research papers on marine invertebrates, which he'd been sending back to London, so impressed the scientists of the day that he was promptly elected a Fellow of the Royal Society. For someone so young and of a lowly upbringing, this was a huge honor.

Huxley was soon lecturing at Imperial College in London (my nemesis—I'm a Queen Mary College alum), where he became friendly with Charles Darwin, another scientist-voyager, Darwin having spent four years as a naturalist on the HMS *Beagle*. When Darwin's *On the Origin of Species* was published in 1859, Huxley became its most vocal promoter in the face of clamorous opposition from the Church of England. The theory of evolution ("survival of the fittest" in biologist Herbert Spenser's quick-and-dirty phrase) directly challenged the Genesis account of our origins. While the church didn't have much to say about how worms and wombats had come about, it was seriously opposed to the notion

Huxley at age 55, 20 years after debating Wilberforce.
LOCK & WHITFIELD

that humans are essentially cousins of (gasp) chimpanzees, gorillas, and orangutans, evolution having done the heavy lifting that, previously, God took care of.

"Would you prefer a monkey for your grandfather or grandmother?" asked Bishop Wilberforce eight months after publication of *Origin*. The occasion was a debate during the 1860 meeting of the British Association for the Advancement of Science between Huxley and Samuel ("Soapy Sam") Wilberforce, Bishop of Oxford. Ironically for the son of William Wilberforce, the abolitionist who had headed the campaign to pass the 1807 (anti-) Slave Trade Act, the younger Wilberforce was a fierce conservative and a formidable debater, "part viper, part soapbox parliamentarian," according to one account.

"The Lord hath delivered him into mine hands," Huxley is supposed to have said to a companion on hearing Wilberforce's taunting question. His response has gone down in history as the finest riposte ever: "If the question is put to me, would I rather have a miserable ape for a grandfather or a man highly endowed by nature and possessed of great means and influence and yet who employs their faculties and that influence for the mere purpose of introducing ridicule into a grave scientific discussion, I unhesitatingly affirm my preference for the ape." *Touché!*

Huxley's adoption of evolution went beyond academic or religious interest. For example, he lectured on behalf of "Negro emancipation," defying the common belief in *polygeny*, the idea that Europeans and Africans are not only different colors but also different species. Polygeny was used as justification for slavery, especially in the American South. As a comparative anatomist nearly a hundred years before Watson and Crick cracked the structure of DNA, Huxley marshaled the available evidence of the time to show that we are all descended from common (black!) ancestors.

Were it that simple. Pro-emancipation Thomas Huxley was a man of his time, an imperialist who believed in racial hierarchies. (Paraphrasing Orwell, we're all equal, but some are more equal than others.) As a result, he became an early eugenicist and one-time president of the British Eugenics Society. The controversy lives on. For example, in 2021, the ultra-woke authorities at Bellingham's Western Washington University removed Huxley's name from its environmental college, citing his "white supremacist values."

24

EVOLUTION'S CODISCOVERER

Crediting Alfred Russel Wallace

July 1 should be the day on which we celebrate the theory of evolution. That's when, in 1858, this revolutionary idea was presented to the world—at least to the august Linnean Society of London—in the form of papers by both Charles Darwin and Alfred Russel Wallace. However, apparently, the momentous event passed by unnoticed. For example, the following May, Linnean Society president Thomas Bell wrote that the previous year had "not been marked by any . . . striking discoveries." So much for (per philosopher Daniel Dennett) "the single best idea anyone has ever had."

The presentation was prompted by a letter delivered to Darwin two weeks earlier from Wallace, then a relatively unknown naturalist who was midway through an eight-year exploration of the East Indies. In it, the younger Wallace laid out the core idea on which Darwin had been working for the previous 20 years: evolution of species by natural selection.

Although Wallace and Darwin received equal billing at the time, Darwin is most associated with evolution, even by its shorthand name: "Darwinism." That's because, though the 1858 presentation came and went with barely a ripple, Darwin's *On the Origin of Species*, published in December 1859, was a bombshell that shook the Victorian world. Yet of the two, it's Wallace who is in many ways the more interesting and controversial character:

- Like Darwin, Wallace was a naturalist-explorer, first in the Amazon Basin and later in what was then known as the Malay Peninsula. Unlike Darwin, who was born into wealth, Wallace struggled financially for much of his life. (Darwin helped secure him a small pension from the U.K. government in 1881.) In contrast to Darwin the materialist, Wallace balked at the idea that evolution could account for higher human mental faculties. He believed that, while evolution was responsible for our physical makeup, consciousness must have a nonmaterial origin, leading to his belief in spiritualism.

"Decent to a fault" (per historian Charles Smith), Alfred Russel Wallace (1823–1913) was a scientist, social justice activist, and polymath. *BORDERLAND* MAGAZINE, PUBLIC DOMAIN

- He opposed eugenics at a time when many prominent evolutionists were advocating it, writing, "Those who succeed in the race for wealth are by no means the best or the most intelligent." During his five years in the Amazon Basin, meeting untold numbers of native people, he adopted the position (rare at the time) that "uncivilized" peoples are neither morally nor intellectually inferior to civilized peoples. Closer to home, he advocated for workers' rights, including a minimum wage and double-time rates for overtime. He was also an ardent supporter of women's suffrage.
- He was a pacifist, writing that "all modern wars are dynastic; that they are caused by the ambition, the interests, the jealousies, and the insatiable greed of power of their rulers, or of the great mercantile and financial classes which have power and influence over their rulers; and that the results of war are never good for the people, who yet bear all its burthens."
- An early environmentalist, Wallace advocated for the protection of rural lands, parks, and "greenbelts" (to prevent the physical growth of urban areas—an idea finally passed into law in the United Kingdom in 1955). In 1887, he spent time in California while on a lecture tour of the United States, visiting redwood groves and the Yosemite Valley with naturalist John Muir.

And much more: delineating between Asian and Australian fauna in the Indonesian archipelago (the "Wallace Line"), advocating the use of paper money, theorizing about color vision, debunking Percival Lowell's mistaken notions about canals on Mars, and evaluating the possibility of life on other worlds. He wrote 22 books and more than 500 scientific papers. While in his eighties, he wrote or edited more than 4,000 pages of scientific papers!

In short, he was (a sobriquet originally applied to Thomas More) a man for all seasons.

25

THE OTHER EVOLUTIONIST

Did a farmer upstage Darwin and Wallace?

Having given Charles Darwin and Alfred Russel Wallace their due, I'd be remiss if I didn't at least acknowledge Patrick Matthew (1790–1874), who some (few) historians say should be given credit as the original discoverer of natural selection, anticipating the other two by nearly 30 years.

In 1831, Matthew published a book in which he uses the term "natural process of selection." However, his contribution, if it can be called that, is that he published his ideas in the appendix to a book on growing trees for shipbuilding, *Naval Timber and Arboriculture*.

He wrote, "As the field of existence is limited and pre-occupied, it is only the hardier, more robust, better suited to circumstance individuals, who are able to struggle forward to maturity, these inhabiting only the situations to which they have superior adaptation and greater power of occupancy than any other kind." This sounds like a neat summing up of the principle, summarized three decades later as "survival of the fittest."

The year after publishing *On the Origin of Species* in 1859, Darwin was made aware of Matthew's earlier book. True to his reputation for honesty and courtesy, he then included a statement in subsequent editions of *Origin*, noting that Matthew had anticipated "precisely the same view on the origin of species" and "clearly saw . . . the full force of the principle of natural selection." Matthew later wrote, "To me the conception of this law of Nature came intuitively as a self-evident fact, almost without an effort of concentrated thought."

So here we have both Darwin and Wallace putting in many dogged and fraught years of fieldwork in South America and (for Wallace) Southeast Asia. Meanwhile, this guy Matthew, quietly working away in apple orchards in Scotland, figures it out "intuitively" because it was "self-evident"!

The late, great evolutionary biologist Ernst Mayr wrote in *The Growth of Biological Thought*, "Patrick Matthew undoubtedly had the right idea, just like Darwin did on September 28, 1838 [Darwin's "aha" moment], but he did not devote the next twenty years to converting it into a cogent theory of evolution. As a result it had no impact whatsoever."

Richard Dawkins sums it up in *Darwin's Five Bridges: The Way to Natural Selection*: "I am increasingly inclined to agree with Matthew that natural selection itself scarcely needed discovering. What needed discovering was the significance of natural selection for the evolution of all life." That is, trees are one thing; life in all its manifestations, 4 billion years of it, is another.

All of this begs the question. In retrospect, the basic mechanism of evolution (heredity, variation, selection) is obvious. Archimedes or Lucretius could have figured it out 2,000 years ago. Why did it take so long for a Scottish apple farmer to realize this "self-evident fact"?

26

DAWN REDWOODS

An overlooked branch of evolution

If a man has to have a memorial or a monument to commemorate his life and his work, nothing could be more fitting than a living Dawn Redwood.

—ANDREW GENZOLI, HUMBOLDT *TIMES-STANDARD*, MARCH 18, 1971, REFERRING TO RALPH WORKS CHANEY

If you're visiting Northern California, Humboldt County's Avenue of the Giants is a must-see attraction. It's also home to a small plaque on a mossy rock at the start of the 2.4-mile-long Drury-Chaney trail, just south of Pepperwood, reminding us of a remarkable story in the history of redwood trees. The plaque reads, "Ralph Works Chaney Memorial Grove. In tribute to a distinguished conservationist, scholar and teacher of paleobotany . . . who found *Metasequoia* (dawn redwood) in China, 1948. President, Save-the-Redwoods League, 1961–1971."

As a backstory, in 1941, while preparations for the attack on Pearl Harbor were underway, a Japanese paleobiologist from the University of Kyoto was engaged in a more benign activity. Dr. Shigeru Miki identified a hitherto unknown species of redwood in fossils from the Mesozoic Era (252 million to 66 million years ago) on Japan's largest island, Honshu. He gave his discovery the genus name *Metasequoia*, soon to be dubbed "dawn redwood." Until then, redwoods were known to come in just two

Left to right: *Sequoia sempervirens* (coast redwood), *Sequoiadendron giganteum* (giant sequoia), *Metasequoia glyptostroboides* (dawn redwood). FROM NATIONAL PARK SERVICE POSTERS, JEDEDIAH SMITH VISITOR CENTER, CALIFORNIA

species, *Sequoia sempervirens* (the coast redwood of Northern California) and *Sequoiadendron giganteum* (the giant sequoia, found mostly south of Yosemite National Park).

Which is where the matter might have rested but for the alert eyes of an employee of the Chinese Bureau of Forestry. While on a field trip to Szechuan Province in central China, Tsang Wang spotted three unusual trees outside the village of Mo-Yao-Chi. Curious, he took specimens to Nanking, where the local botany experts were unable to identify them. Finally, the director of Beijing's Fan Memorial Institute of Biology realized that these were living specimens of Miki's fossils! That was in 1946.

When the news reached paleontologist Ralph Chaney, then a 58-year-old professor at the University of California, Berkeley, he decided to see the trees for himself. It helped that he was an old China hand. By the time of his death in 1971, he'd made nine trips there, including the legendary 1925 American Museum of Natural History

Not extinct! Dawn redwood outside the Eureka, California, City Hall, cultivated from seedlings obtained by Dr. Ralph Chaney in China in 1948. AUTHOR dinosaur expedition to Mongolia led by Roy Chapman Andrews, allegedly the model for Indiana Jones.

Backed by the Save the Redwoods League, he flew to Chongquing in southwest China in February 1948 accompanied by *San Francisco Chronicle* science editor Milton Silverman. Although the Chinese Civil War was raging, the pair boated from Chongquing down the Yangtze River and trekked across three mountain ranges to Mo-Yao-Chi and thence to the nearby Valley of the Tiger, where they collected seedlings and cones from scores of dawn redwoods. After surviving a bandit attack on their return trek, then convincing U.S. customs in Hawaii that he was importing "antiques" (since live plants were prohibited), Chaney spent years cultivating his precious trophies in a hothouse on the Berkeley campus before sending tens of thousands of seedlings to many parts of the United States.

Dawn redwoods differ from the coast redwoods you'll see on the Avenue of the Giants since its branches go up, not out, with a distinctive bark. If you'd like to see all three redwood species—coast, giant, and dawn—in one place, check out the parking lot of the Jedediah Smith Visitor Center on State Highway 199 north of Crescent City, California. For a tree that was unknown until 1941 (and then only from ancient fossil specimens), the dawn redwood there sure looks full of *joie de vivre* today.

EXPLORATION

To the ends of the Earth

27

FIFTH-CENTURY PACIFIC CROSSING?

Evaluating the evidence

We know that Columbus wasn't the first non-native to visit the Americas since we have hard evidence that Vikings, under Leif Ericson, created an exploration base on the Atlantic shore of Canada around AD 1000. "L'Anse aux Meadows," situated on the northernmost tip of Newfoundland, is evidence of the earliest *incontrovertible* pre-Columbian discovery.

Which leaves a host of *controvertible* discoveries. Over the years, discovery of the Americas from the east has been credited to, among others, Egyptians, Phoenicians, Africans, Trojans, Carthaginians, Romans, Arabs, Irish, Welsh, Germans, Poles, and at least one of the Ten Lost Tribes of Israel, all of whom are supposed to have crossed the Atlantic Ocean well before 1492.

One story with slightly more credibility than the others and coming from the other direction—across the Pacific—is that of the Buddhist missionary Hui Shen, who, as an old man, visited the Chinese emperor Wu Ti at his court in AD 499. He claimed that 40 years earlier, he and four companions sailed to a place named Fusang, 20,000 *li*, or 5,000 miles, east of China. If the story is to be believed, he landed in present-day British Columbia, as shown on the accompanying French map of 1792: "Fousang des Chinois" (arrow).

According to the account in the seventh-century *Book of Liang*, Hui Shen told the emperor, "The native tribes in Fusang were civilized, living

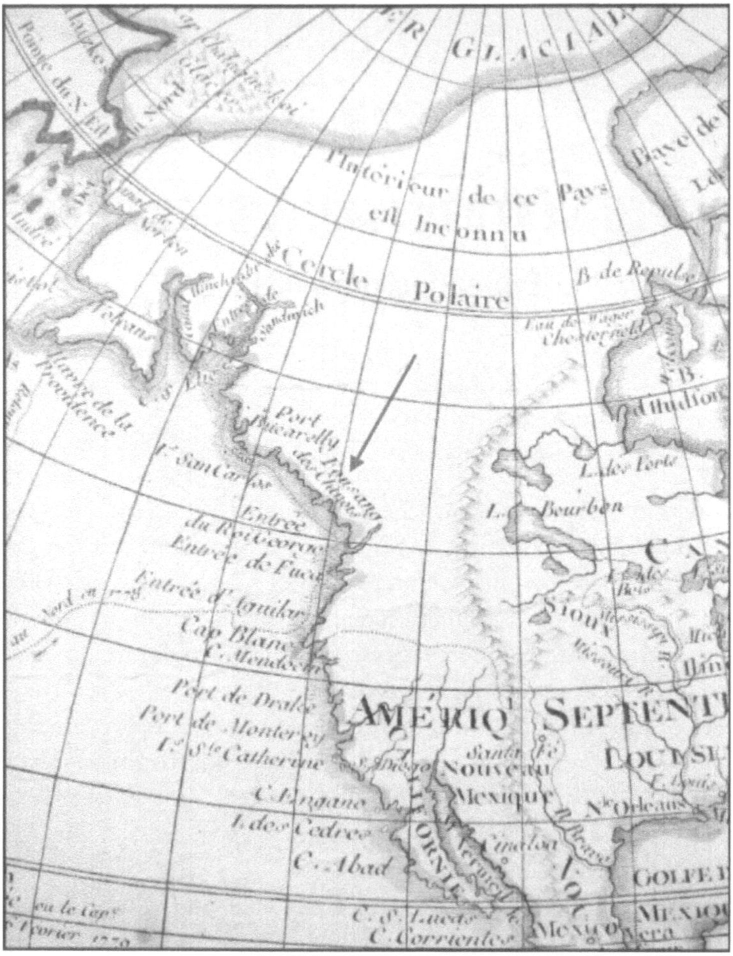

French map of 1792 showing "Fousang des Chinois" (arrow), supposedly the part of British Columbia visited by Hui Shen in the fifth century. PUBLIC DOMAIN

in well-organized communities. They produced paper from the bark of the Fusang plant [mulberry? maguey?] for writing and produced cloth from the fibers of the bark, which they used for robes. . . . A main chief, with the help of several officials, governed the country. The majority of people were law-abiding citizens. . . . The marriage arrangement was relatively simple. If a boy wanted to marry a girl, he had to build a cabin next to the home of the girl and stay there for a year. If the girl liked him they would get married; otherwise he would be asked to go away."

Here are some credibility points for the story:

- We know that Chinese shipwrights were building oceangoing junks as early as the second century AD.
- To prove that a transpacific voyage was possible, five young Chinese men sailed a junk from Taiwan to San Francisco in 1955. They came the direct way across the open ocean. If you follow the coastline, the longest stretch of open sea between China and North America is only 200 miles, between the Komandorski and the Aleutian islands.
- The Maya people of Central America did write hieroglyphics on bark paper.
- A newspaper account from 1882 claims that a miner found 30 Chinese coins (since lost!) 25 feet below the surface near the old asbestos town of Cassiar, north of Dease Lake, British Columbia. According to a the report, "The coins were dated to 1200 BC."

Beyond these slim clues, we really can't say one way or another if Hui Shen—the "Leif Ericson of China" according to Alexander von Humboldt—made it to the Americas, the land he called Fusang. What would clinch the tale would be the discovery of a wrecked Chinese junk off the Pacific coast.

Postscript: One of my proofreaders asked, *If wrecked, how did Hui Shen return?*

28

THE UNITED STATES OF ZHENG

Showing the flag 600 years ago

The name "America" comes from Amerigo Vespucci, the Italian explorer who first claimed that Brazil and the West Indies are parts of a separate continent, not, as Columbus believed (against all evidence), part of Asia. But how close did those of us in the United States come to be living instead in the "United States of Zheng"? *Really* close, according to ex–British Royal Navy officer Gavin Menzies, author of the 2003 bestseller *1421: The Year China Discovered America*. As we'll see, his is strictly a minority view.

Menzies' assertion is based on the voyages of the legendary Chinese eunuch-admiral Zheng He (1371–1433), who made seven long voyages starting in 1405, commanding huge fleets that included the largest wooden ships of all time. His first two-year voyage sailed from China to the southern coast of India via present-day Vietnam, Java, Sumatra, and Sri Lanka. According to the surviving records, his fleet consisted of 317 ships manned by more than 25,000 crew. In six subsequent voyages, his fleet sailed as far as Aden (in Yemen) and down the east coast of Africa to Mombasa (Kenya).

These were not voyages of discovery since they all followed long-established and well-mapped trade routes. Rather, they were designed to "show the flag" and impress on foreign rulers the symbolic authority of the Chinese Ming emperor. I bet they did too! No other country in the world could field such an imposing fleet as Zheng He's, including four-deck, nine-masted "treasure ships," each the size of a football field.

Model of a huge Ming-dynasty Chinese "treasure ship" compared to one of Columbus's ships in the Ibn Battuta Mall, Dubai. LARS PLOUGMANN, CREATIVE COMMONS

The voyages abruptly stopped following the death of 62-year-old Zheng He in 1433, homeward bound from his seventh expedition. Probably a combination of court intrigues and financial constraints was to blame. Zheng He had been part of an influential eunuch faction in the emperor's court, opposed by the more conservative Confucian bureaucrats. With Zheng He dead and costly Mongol raids threatening the northern frontier, Chinese emphasis shifted from the sea to the land, and xenophobic scholars destroyed most of Zheng He's records.

But ah, the what-ifs of history! If Lee Harvey Oswald had missed, would Kennedy have pulled out of Vietnam before the war escalated? If an unusually fierce storm hadn't wrecked 60 ships of the Spanish Armada in 1588, would Brits (and Americans) be speaking Spanish? And if Zheng He had sailed east, across the Pacific, would you be reading this

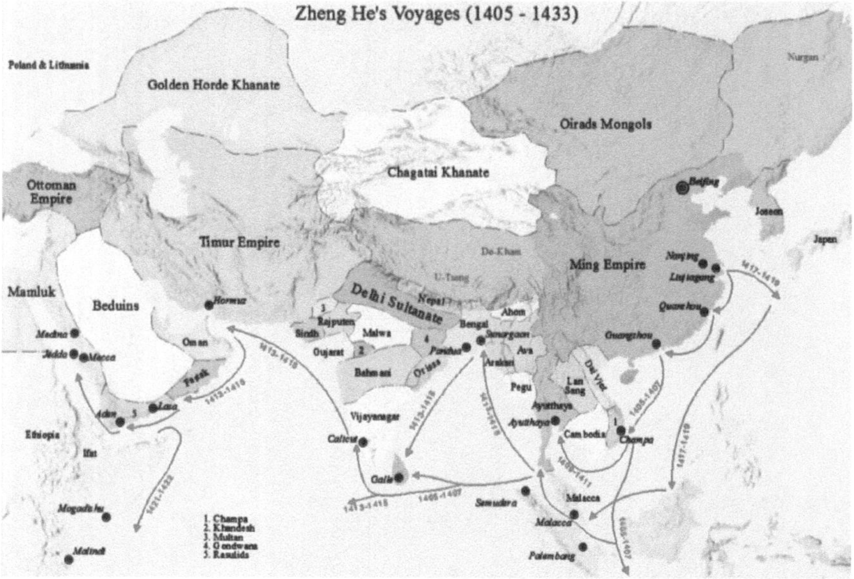

Voyages of Zheng He during the Ming dynasty. "SY," CREATIVE COMMONS VIA WIKIPEDIA

story in Chinese logograms (all 7,000 of them)? Would we be living in the United States of Zheng?

That's where Gavin Menzies' theory comes in. In his rewriting of history, Zheng He's fleet not only "discovered" what we call America but also circumnavigated the globe, one-upping Columbus, Vespucci, Magellan, and all those European Johnny-come-latelies. In 550 pages, Menzies cites a host of dubious sources, including, for example, a pre-Columbian Chinese map showing Australia ("unfortunately . . . lost") and Chinese anchors found off the coast of California (absent documentation). He contradicts himself (claiming, for example, that fifteenth-century sea levels were both higher and lower than now) and does pretty much everything to raise the hackles of professional historians, his forte.

It's a great story, but, as far as I know, no reputable researcher has signed on to his theories. Still, Menzies sold a million copies of *1421*, and his subsequent book on Atlantis isn't doing badly either. Me, I'll stick to evidence-based history.

Unless, of course, one of those six-figure advances comes my way.

29

THE MARTELLUS MAP

Was Columbus misled?

In February 1970, two days after arriving in Canada as an immigrant, I was sitting in an employment office in Vancouver. My interviewer said, "There's an opening at 'Indian Affairs.' Would you be interested?" Sure, I thought, naively anticipating that I'd be working with Hindus. I realized only later that the Indians in question were the folks now mostly referred to as Native, Indigenous, or First Nations people. Blame Christopher Columbus for my confusion, although the indignity can be traced back to early mapmakers, including a polymath who lived in Alexandria, Egypt, in the second century.

PTOLEMY

Claudius Ptolemy was a Greco-Egyptian mathematician, astronomer, and geographer whose mapmaking skills were used by navigators for more than 1,000 years after his death in AD 170. His greatest surviving accomplishment is the *Geographia*, in which he combined maps of the known world, including a gazetteer of 8,000 places listed by latitude and longitude (a first), together with a treatise on cartography, including, crucially, how to project Earth's curved surface onto a flat piece of paper.

By the time his work resurfaced in Constantinople around 1300, the original maps had been lost, but Renaissance cartographers were able to reconstruct them from his gazetteer. In looking at the results, it's obvi-

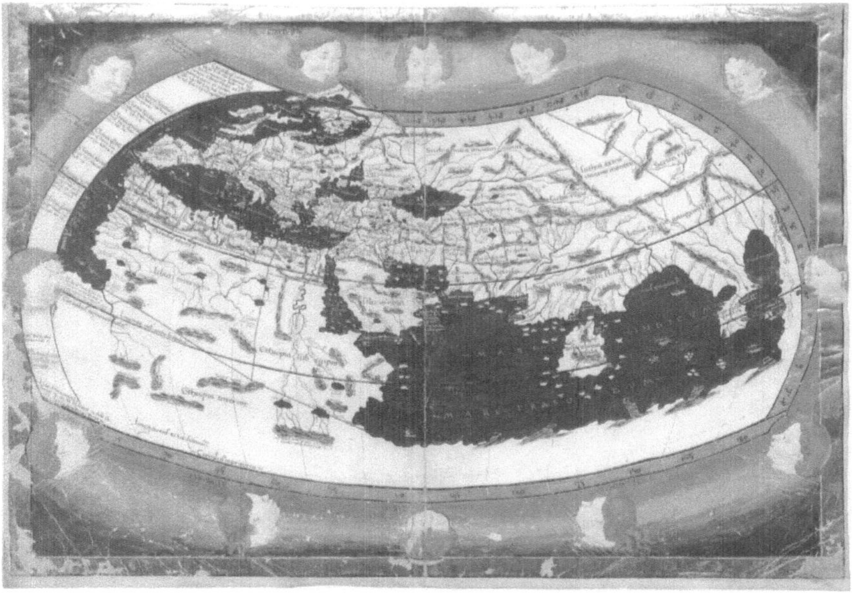

1467 map by Jacob d'Angelo, based on Ptolemy's *Geographia*, showing lines of latitude and longitude. Note that the coast of China is cut off. WIKIMEDIA COMMONS

ous that Ptolemy greatly overestimated the width of Eurasia. His map showed the longitudinal width between Britain in the west and a vaguely outlined China in the east spanning 180 degrees; the true value is about 130 degrees. Only in 1295, when Marco Polo returned from his astonishing travels in Asia, was "Cipangu"—Japan—made known to Europeans.

One of the first charts to incorporate Polo's knowledge was Hendricus Martellus's four- by six-foot world map of 1491. The problem was that Polo's chronicles gave very vague descriptions of geography ("this city is five days camel journey east of that one"), so places are only loosely located, including Japan, which he'd only been told about. However, in addition to locating Japan *much* farther east than it actually is, Martellus's was the first post-Ptolemy map to show latitude and longitude.

COLUMBUS

In theory, this should have made it possible for Christopher Columbus to figure out the distance from the Canary Isles to Japan in his quest to find a westerly route to the "Indies." Except it didn't. Basically, Columbus *way* underestimated the size of our planet, adopting the equivalent of 18,800 miles for Earth's circumference instead of the actual value of 25,000 miles, confusing "Arabic" and "Roman" standard units. Or—historians disagree on this—he may have known the true value but chose to go along with a smaller one. Why? Because funding for his proposed transatlantic voyage depended on his backers believing that the distance from the Canary Islands to Japan was no more than 2,300 of our miles. It's actually 7,800 miles! Whatever doubts he may have had, he kept them to himself when he approached the rulers of first Portugal and then Spain for funds.

Fortunately for him but not so fortunately for the natives he encountered, his expedition made landfall in the Bahamas off the coast of what would soon be called "America" (after the Italian navigator Amerigo Vespucci). Still, Columbus thought he'd found the Indies (as in India), the name then given to all the lands of eastern Asia—so he called the local populace "Indians." (An alternative theory—that "Indians," Spanish *Indios*, came from *la gente en Dios*, "the people of God"—is almost certainly wrong.) Even after Columbus was shown evidence of his mistake, he persisted in his belief, which is why we still have places like "Indiana" and the "West Indies" and why both the United States and Canada have many "Indian Acts" on their books. Thankfully, referring to a Native American as an "Indian" is going the way of "Negro" and other unenlightened designations.

THE MAP

Despite its importance to Columbus and other navigators of the late fifteenth and early sixteenth centuries, the 1491 Martellus map has also been one of the most frustrating for historians. You can see the problem on the left side of the illustration, where the text has faded to the point of near invisibility. About the time Yale University acquired the map in 1960, crude (by today's standards) ultraviolet imaging showed that there was much more to the map than met the naked eye. In August 2014, an intense 10-day program using modern multispectral imaging techniques,

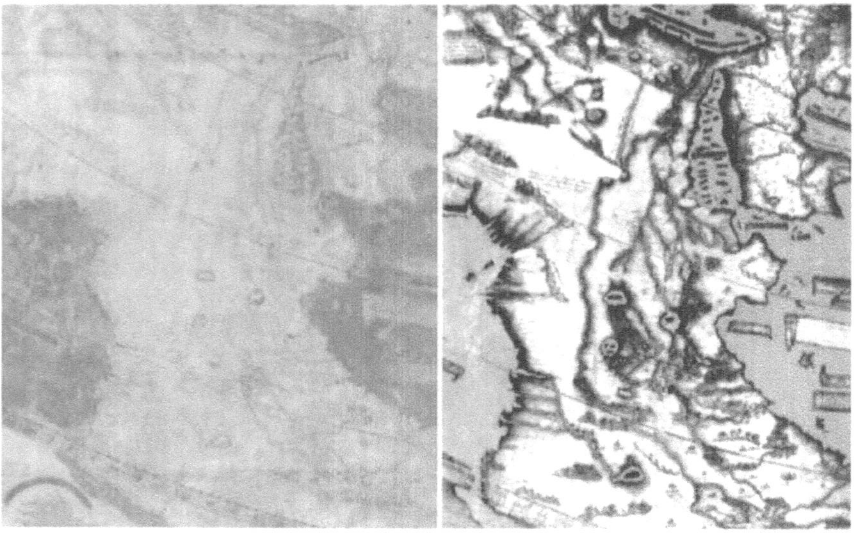

The Arabian Peninsula in Martellus's 1491 map: visible light (left) and multispectral imaging (right). This map supported Columbus's contention that Japan was only 90 degrees west of Lisbon, less than half the actual distance. PUBLIC DOMAIN

appropriately named the Lazarus Project, resulted in the detail shown on the right side.

The researchers photographed the map in 12 wavelengths of light from ultraviolet to infrared, ending up with 12 terabytes of data, which they manipulated to reveal previously unseen details. (Four terabytes is roughly half of Wikipedia's primary SQL database.) If you've played with high-dynamic-range photography, you'll have a rough idea of what's involved. The researchers' software identified the best combination of wavelengths for each section of the map. After weeks of massaging data, virtually all the text and geographical features became visible, and the map could finally be studied in detail.

The project turned out to be a bonanza for historians, laying bare hundreds of place-names and descriptions, resulting in a much clearer understanding of Renaissance cartography. According to the project leader, historian Chet Van Duzer, the map is a "missing link in our understanding of people's conception of the world." The new analysis shows how much Martellus relied on Ptolemy's ancient *Geographia* from about AD 150 together with information brought back from China by Marco Polo and the 1488 discovery of the Cape of Good Hope by Portuguese explorer Bartholomew Diaz.

The influence of Martellus's map reached far and wide. Martin Behaim of Nuremburg, Germany, drew heavily on it in preparing the oldest extant globe in 1492. So did Martin Waldseemüller, the German cartographer who first called the newly discovered continents "America" on his 1507 map, the "birth certificate of America." Which is why, in 2003, the Library of Congress bought it for $10 million.

30

MAGELLAN'S PACIFIC

Linking the oceans

In the course of the voyage, Magellan had managed to outwit death many times. He overcame natural hazards ranging from storms to scurvy, and human hazards in the form of mutinies. In the end, the only peril he could not survive was the greatest of all: himself.

—LAURENCE BERGREEN, *OVER THE EDGE OF THE WORLD: MAGELLAN'S TERRIFYING CIRCUMNAVIGATION OF THE GLOBE*

Although Ferdinand Magellan (1480–1521) was Portuguese (Fernão de Magalhães), he was actually in the service of the king of Spain, Charles V, when he gave the Pacific Ocean its name. Previously, Magellan had suffered a rather messy relationship with his native Portugal. Although he helped install a Portuguese viceroy in India, he was also accused (probably justly) of going AWOL and subsequently illegally trading with the Moors.

In Magellan's time, Spain had a problem: the only known shipping route to the Indies (roughly present-day Indonesia) was via the Cape of Good Hope. The right to that passage had—unfortunately for Spain—been granted to Portugal under the 1494 Treaty of Tordesillas, in which Pope Julius II gave his blessing to dividing the world into Spanish and Portuguese spheres of influence. So the enterprising Magellan proposed

Arrows indicate the route of the Strait of Magellan, which joins the Atlantic and Pacific. (GNU LICENSE) Inset: Ferdinand Magellan. LIBRARY OF CONGRESS

to Charles V that he would find an alternative westward route, sailing south and around the newly discovered continent of South America.

Charles agreed to Magellan's ambitious proposal, giving him five ships. The small armada sailed downriver from Seville in August 1519 to begin what turned out to be an ill-fated journey. Among other trifles, Magellan put down a mutiny in Patagonia, tortured the mutineers, and then lost two ships—one wrecked, one deserted. Finally, three of his original five ships became the first to sail through what is now called the Strait of Magellan in the "single greatest feat in the history of maritime exploration" (Laurence Bergreen). It took 38 days to navigate the 300-mile channel from the Atlantic to the Pacific between Tierra del Fuego and the mainland before emerging into the Pacific on November 28, 1520. In gratitude for finding relatively calm seas after the rough passage, Magellan named the new ocean Mar Pacifico.

He didn't have long to savor his triumph, losing his life six months later during a totally unnecessary skirmish with native people in what is now the Philippines. Of 237 men who set out in five ships to sail around Earth in 1519, only 18 emaciated crewmen completed the first global circumnavigation, limping home to Spain in September 1522 on Magellan's flagship *Trinidad* without a captain, officers, or boatswains.

The name of the ocean stuck, though, and I like to think that Magellan's ghost would be happy to know he was responsible for naming the largest body of water on Earth.

31

COLUMBUS OF THE PACIFIC

Friar Urdaneta's "Volta do Mar"

In 1565, 73 years after Columbus navigated across the Atlantic, Basque friar Andrés de Urdaneta (1508–1568) discovered a vital route from Manila in the "Old World" to Acapulco in the "New World." His discovery opened up Spanish colonization of the Philippine archipelago by establishing the "Manila galleon" trade route linking the spice-rich islands with the port of Acapulco in the viceroyalty of New Spain (present-day Mexico) for the next 250 years.

As a monk in the Order of the Hermits of Saint Augustine, probably the last thing 56-year-old Urdaneta had on his mind was leading an expedition across the Pacific in response to a plea from King Philip II of Spain. But in 1564, Urdaneta probably knew more about sailing the world's largest ocean than anyone. He'd already had his share of adventures, being part of the ill-fated Loaísa expedition, in which just 79 men survived out of the original crew of 450 on seven ships. Urdaneta didn't return home for 11 years, spending several years in captivity before being brought back under guard to Lisbon by the Portuguese "Indian Armada" (in the process making the second circumnavigation of the globe). He subsequently escaped from Portugal and, after receiving a cool reception from the king of Spain, decided he'd had enough of Europe, so he headed to New Spain for what he probably imagined would be a life of holy contemplation through his twilight years.

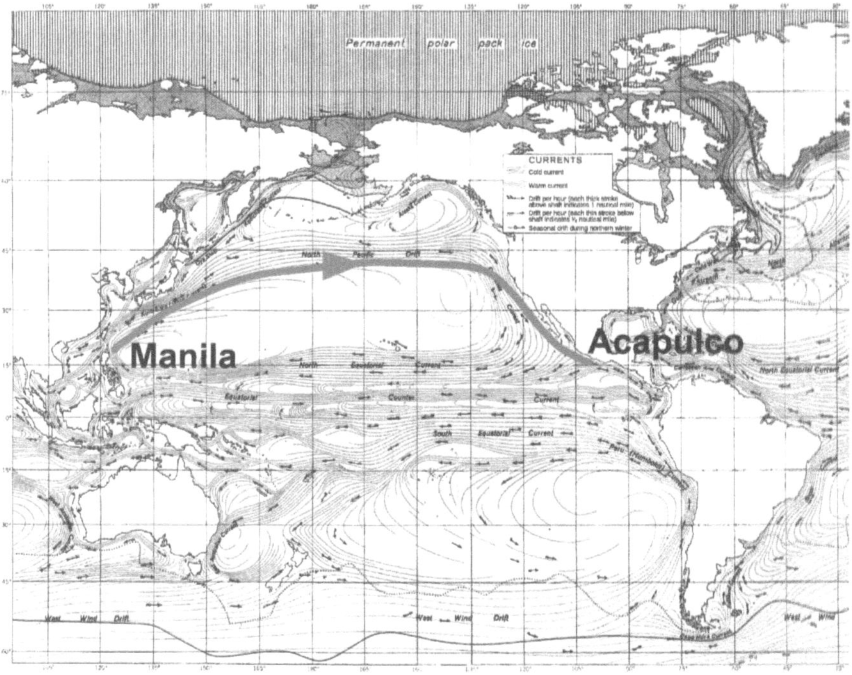

Urdaneta's return route, or *tornaviaje*, from Spain's Philippine colony to New Spain (now Mexico). U.S. ARMY 1943/WIKIPEDIA COMMONS

The challenge thrown down by the new Spanish king, Philip II, was simple: find an easterly route across the Pacific for galleons to safely sail from Spain's newly acquired Philippines colony to Acapulco. Sailing *westward* across the Pacific was no great problem; Magellan had been blown by the westerly trade winds across the Pacific in 1521. The *eastward* voyage proved trickier. Several Spanish navigators had already failed to find suitable winds before Urdaneta established the *Volta do Mar* Manila-to-Acapulco route in 1565.

He guessed correctly that, by sailing sufficiently far north, he would leave the easterly trade winds and enter the zone of the westerlies. Setting sail from present-day Cebu Island on June 1, 1565, he sailed north to beyond 35 degrees latitude, where he found westerly "north Pacific drift" winds. One hundred and thirty days and 12,000 miles later, he reached

Acapulco after a terrible voyage. Fourteen of his crew died from scurvy (what we now know to have been a deficiency of ascorbic acid, vitamin C). Arriving at Acapulco, only Urdaneta and one other seaman had strength enough to drop anchor, but it was enough to ensure that "Urdaneta's route" could be used by future generations of galleons between the Old and New Worlds.

The first land sighted by Urdaneta on his groundbreaking (seabreaking?) voyage of 1565 this side of the Pacific was almost certainly Cape Mendocino, the most westerly point on the coast of California. The headland had been named 20 years earlier by its Spanish discoverer, Juan Rodríguez Cabrillo, to honor the first viceroy of New Spain, Don Antonio de Mendoza. Now the landmark was to be a vital point of navigation for trade galleons following Urdaneta's passage en route to Acapulco. Urdaneta died three years later at a friary in Mexico City, an old man by the standards of his time.

With the route established, every year from 1565 to 1815, between two and four large Spanish galleons, built from Philippine hardwood, made four- to six-month eastbound voyages from Manila to Acapulco, bringing rich cargoes of porcelain, ivory, silks, wax, chinaware, and spices. Finally, the Mexican War of Independence put a stop to the trade in 1815, six years before Mexico finally seceded from Spain after a long and bloody struggle.

KING PEAK TREASURE TROVE?

Of the hundreds of "Manila galleons" that sailed from the Philippines to Acapulco over 250 years, 30 were lost without a trace, some of which probably ended up on California's north coast, which could explain the "Legend of King Peak." In the July 1963 issue of *Western Folklore* magazine, Humboldt County author Lynwood Carranco retold a tale he'd heard from *Humboldt Times* columnist Andrew Genzoli. Genzoli claimed that as a youngster, he was told the following from Johnny Jack, a Mattole and Wiyot tribal member who lived at the mouth of the Mattole River. A ship carrying a rich cargo of gold, gems, and silk was wrecked on Northern California's Lost Coast. The crew either drowned or were killed by local Native Americans who recovered the cargo and stashed it

Model of the type of Spanish galleon used on the Manila-to-Acapulco trade route, in the Acapulco Fort museum, Mexico. AUTHOR

in a cave below King Peak, the tallest mountain on California's rugged Lost Coast. An earthquake subsequently sealed up the opening of the cave, and (you have to take my word on this) the loot is there to this day. Does this tale, passed down through the years by members of the local Mattole tribe, explain the disappearance of one of the missing Manila-to-Acapulco galleons?

The truth is out there!

32

THE LONGITUDE CHALLENGE

Latitude, easy. Longitude, not so much

For nearly 2,000 years, since the time of Greek geographer Claudius Ptolemy, travelers and mapmakers have defined a position on Earth by two numbers. *Latitude* tells you how far you are from the equator, measured in degrees, 0 to 90, north and south. *Longitude* is your angular distance from an imaginary "meridian" line drawn from pole to pole, running through the Royal Observatory, Greenwich, outside London, England. It's also measured in degrees: 180 degrees west of Greenwich and 180 degrees east for the full 360 degrees. (Why 360? Blame the Babylonians, who found it an easy number to divide into.)

It's comparatively easy to reckon one's latitude. Earth's axis points to two imaginary "celestial poles" in the sky. North of the equator, the north celestial pole is currently less than one degree from the North Star, which is easy to find. Measure the angle from your horizon to that virtual point in the sky, and you've got your latitude. (It's zero at the equator, where the north celestial pole is on the horizon, and 90 degrees at the North Pole, where it's directly overhead.)

Longitude is a much trickier proposition. Consider a point on Earth where local noon (that is, the sun at its zenith) is exactly eight hours later than at Greenwich; Lake Tahoe, for example, fits the bill. Eight hours is one-third of a full day, which converts to one-third of the full 360 degrees; that is, Lake Tahoe is at longitude 120 degrees west. So if you know the time difference between your local noon and the local noon at

Greenwich, you've got your longitude. The challenge, then, that navigators have wrestled with for centuries can be simply stated: what time is it at Greenwich? Today, GPS satellites give anyone with a smartphone that information. Formerly, radio signals, chronometers, and the stars were used. I'll be focusing on the last one.

- Radio: The U.S. Navy began broadcasting "Greenwich time" signals in 1905, and the rest of the world quickly followed. Earlier—in 1872—surveyor John Wesley Powell used a telegraph line to confirm his longitude near the Grand Canyon using time signals received nearly instantly from Salt Lake City, where local clocks kept track of Greenwich time.
- Chronometers: Before radio, navigators carried accurate timekeeping devices set to Greenwich time. John Harrison's 1759 marine chronometer, dubbed "H4," solved the wickedly difficult problem of keeping accurate time on a rolling ship at sea. As anyone who drives a car or uses a washing machine knows, mechanical devices are prone to failure, hence the navigators' adage, "Never go to sea

John Harrison's H4 marine chronometer, or "sea watch." MIKE PEEL, CREATIVE COMMONS with two chronometers, take one or three." If you have three and one shows a different time from the other two, ignore it. (If you have two and they show different times, which one do you trust?) Marine chronometers—HMS *Beagle* carried 22—were hugely expensive and impractical for land explorers such as Lewis and Clark. (Lewis had obtained an English chronometer in Philadelphia, but it wasn't reliable when traveling over land for long periods, and he just used it as a watch to connect the time periods between astronomical sightings taken over a few days.)

- Celestial navigation: As long ago as 150 BC, Hipparchus of Nicea wrote that lunar eclipses (which occur at roughly the same time for all Earth-bound observers) could be used to establish longitude. Of the many "clocks" in the night sky, some of the most accurate are eclipses of Jupiter's four bright moons. Galileo, who discovered these moons in 1610, realized that an eclipse of, say, Io could be clocked simultaneously by observers at two places on Earth, thus giving their relative longitude. This was the breakthrough that changed navigation. Soon after, in 1668, Italian astronomer Giovanni Domenico Cassini printed the first tables of Galilean eclipses. Later, now working in the Paris Observatory, he published the *Connaisance de Temps*, an almanac (strictly, an emphemeris) predicting the exact times of the every-other-day eclipses of Io as viewed from a meridian running through Paris. If you had a telescope, carrying his almanac in your pocket was like being in instant communication with a home-base observer.

Within a few years of publication, cartographers used Cassini's *Connaisance de Temps* to redraw the map of Europe that had been mostly unchanged since Roman times. On seeing the new map, the story is that the French king complained that his mapmakers had stripped more land from France than had all the wars fought by his predecessors. Not to be outdone, the Royal Observatory in Britain began issuing its own *Nautical Almanac* in 1766, stating that its predictions for eclipses of the Galilean moons "are well known to afford the readiest and . . . best method of settling the longitudes of places."

Compare Abraham Ortelius's 1570 map of Europe (top) with that of Robert Janvier (1794), who used Cassini's tables of the eclipse times of Jupiter's moons to determine longitudes.

PUBLIC DOMAIN

MACKENZIE AND VANCOUVER

All this is fine if you have a decent 40-power telescope to fix the exact time of immersion (disappearance behind Jupiter) or emersion (reappearance) of one of the Galilean moons. Alexander Mackenzie, who in 1793 became the first white explorer to complete an east-to-west traverse of the American continent, schlepped a heavy London-made brass telescope. On reaching Dean Channel on the coast of (what is now) British Columbia and having used a sextant to set his watch to the local noon, that night he was able to time emersions of Ganymede and Io with his telescope. He thus determined his longitude by comparing his timing with "Greenwich time" of the same events listed in his *Nautical Almanac.*

Mackenzie just missed meeting Captain George Vancouver, who had explored Dean Channel just six weeks earlier. Vancouver, a skilled navigator, used both marine chronometers and "lunar distances" (another celestial navigation trick) to determine longitudes. Vancouver determined the latitudes and longitudes of prominent features along the Pacific coast, including Cape Disappointment on the north side of the mouth of the Columbia River, near where Meriwether Lewis and William Clark wintered 12 years later (1804–1805).

LEWIS AND CLARK

Lewis and Clark were army men, not trained surveyors, and their attempts at fixing locations along their route were mostly unsuccessful. Clark's map (published in 1814, a full eight years after their return) was anchored not on data obtained during the expedition but on three previously known points: their starting point, St. Louis, the approximate midway point at Fort Mandan, whose location had been fixed by professional surveyor David Thompson in 1797 using lunar distances, and Cape Disappointment. Clark mapped the rest by dead-reckoning, which explains, for example, his 50 percent overestimate of the width of the Rockies.

Today, my iPhone receives GPS data from satellites 12,500 miles overhead, giving—instantly—my location to within about 20 feet. What would those doughty navigators of old have given for such convenience?

33

SCOTT OF THE ANTARCTIC

Reckless and/or brave?

Growing up in Britain in the 1950s during the post–World War II period, when my native country was hungry for inspiration, I learned about the heroes of what was still, erroneously, called the "Empire." (It had morphed into the "Commonwealth" since 1931.) One of the most prominent of these was Robert Falcon Scott (1868–1912), better known as "Scott of the Antarctic." Scott vied with Norwegian explorer Roald Amundsen to win the race to the South Pole, but when Scott's five-man expedition arrived there on January 17, 1912, they realized they'd been beaten by five weeks. While Amundsen and his men were the first to reach the South Pole, most of the glory went to Scott. Although he and all his men perished of exhaustion and starvation on the 800-mile return journey, Scott's "Message to the Public," found a few months later by a search party, became the stuff of legend. In this poignant document, Scott's version of the fate of his expedition claimed that he and his men died not because of any lack of leadership ability but due to chance and bad luck.

This heroic version of Scott's saga was current when I was a schoolboy. Scott's standing as a brave but prudent leader, fated by circumstances beyond his control, kept him in the public eye as the very model of a British gentleman-explorer. True, a few rumblings occasionally challenged that unstintingly positive assessment, but in general, his reputation was

Scott's expedition at the South Pole, January 18, 1912, one day after they discovered that Amundsen had beaten them. Left to right: Lawrence Oates, Henry Bowers (pulling the string to operate the camera shutter), Scott, Dr. Edward Wilson, and Edgar Evans. Scott later wrote, "We are showing that Englishmen can still die with a bold spirit, fighting it out to the end." PUBLIC DOMAIN

secure—until the 1979 publication of a joint biography of Scott and Amundsen by Cambridge historian Roland Huntford, who concluded that Scott was a "heroic bungler" and "one of the worst polar explorers."

Both sides can make a good case. Almost certainly, Scott did a terrible job of planning his return journey from the South Pole by mismanaging the location of food depots (he and the remaining two survivors were only 11 miles from "One Ton" depot when they died but 24 miles *beyond* the original intended location), and, unlike the far-better-prepared Amundsen, Scott failed to take advantage of the tremendous strength and endurance of huskies as sled dogs. On the other hand, a new analysis of weather conditions for the two months following his arrival at the South Pole determined that the weather on the "Barrier" through which the party had trekked was about as bad as it ever gets, so perhaps Scott was more unlucky than incompetent. The argument continues to this day.

Scott's reputation may have more to do with what historian Stephanie Barczewski calls "current cultural forces." That is, 1912 was another world, a time when the public had different expectations for its heroes. Back then, a tendency to recklessness—which probably does apply to Scott—might have been seen as more of a positive attribute than is the case today. Perhaps Scott was both a champion and a bungler, depending on how you look at it. For this British schoolboy, Scott's reputation is safe: he's my hero of the Antarctic and always will be.

HEALTH

Our bodies, ourselves

34

HUNTER VERSUS FARMER AND ADHD

Leave those kids alone

Attention-deficit hyperactivity disorder, or ADHD, is a psychiatric disorder affecting between 3 and 5 percent of children worldwide, many of whom continue to have the symptoms into adulthood. An estimated 8 million adults in the United States suffer from ADHD, which manifests primarily as inattention, hyperactivity, and impulsivity. This, at least, is the official view of the National Institutes of Health and of the psychiatric "bible," the *Diagnostic and Statistical Manual of Mental Disorders*. Meanwhile, arguments rage between researchers, some of whom reject its classification as a disorder and others who believe the condition is real and genetic and who say that twin studies support their point of view.

One of the more controversial approaches to ADHD is the "hunter versus farmer" theory, which holds that the disorder is indeed genetic but, far from being a disadvantage, was an evolutionary asset thousands of years ago when our ancestors were nomadic hunter-gatherers. The idea is that many of the typical traits attributed to ADHD, including impulsivity and novelty seeking, were beneficial at a time when our forebears were struggling to survive in harsh and uncertain pre-agricultural, subsistence environments.

Researchers tested the theory using data from the northern Kenyan Ariaal tribe, which had been 100 percent nomadic until about 70 years ago, when the tribe essentially split in two when some members settled down into an agricultural way of life. Both factions of the tribe were tested for the DRD4 7R+ allele, a genetic variation linked to ADHD symptoms. The study found that the still-nomadic members who had the variant allele were generally healthier than those who lacked it. Conversely, the agriculturalists exhibited the opposite effect: those with the allele associated with ADHD fared worse than those lacking it, mirroring the situation in the United States.

The theory, then, is that ADHD behavior favors nomadic hunters, for whom a short attention span and unpredictable behavior might make the difference between survival and starvation. Significantly, perhaps, contemporary males worldwide—the traditional hunters—are diagnosed with ADHD about three times as frequently as girls. This suggests the possibility that a once beneficial genetic variant is now a maladaptation, similar to our predilection for ingesting as much fat, sugar, and starch as we can get our hands on. In that case, the trait that helped our ancestors survive is today leading to a global epidemic of obesity.

None of this is saying that ADHD is a good thing any more than our genetic disposition to stuff ourselves with Big Macs is healthy. What it does suggest, if the "hunter versus farmer" research stands up to further investigation, is that in order to thrive, some kids need to be taught in a less structured environment, more like that of our nomadic ancestors, than the confining and disciplined classrooms in which they find themselves today.

35

LACTOSE TOLERANCE

Evolution in action

You're standing in front of the dairy section at your local supermarket deciding whether to buy 2 percent or whole milk, at which point you might consider that, globally, adult milk drinkers are in the minority. Most of the world's non-infant population is lactose intolerant, while a minority of us have a mutant gene that allows us to digest milk. Assuming you're in that minority—most of us in the West are—you can thank evolution.

To recap (see "Genetic versus Cultural Evolution"), evolution results from random mutations in the process of passing the genome (the genetic information carried in the egg or sperm) from one generation to the next. Most of these mutations are either neutral or harmful, but a few are beneficial; that is, the bodies that they helped create are better adapted to their environment than are their siblings who lack the mutation. Thus, their genes are more likely to be passed on to future generations. Such is the automatic mechanism of natural selection.

Although the human genome varies by less than half a percent throughout our species, the differences are important. Most of the variances appeared in the past 50,000 years, after most of our common ancestors moved out of Africa to begin the global human diaspora. One of the more obvious differences is *lactose tolerance*, which appeared a mere 400 generations, say, 10,000 years, ago.

Aurochs (*Bos primigenius*), ancient relatives of modern cattle, were once herded for milk. Depicted in many cave paintings, aurochs were huge: adults were six and a half feet at the shoulders and weighed 2,200 pounds. They went extinct about 400 years ago. PUBLIC DOMAIN VIA WIKIPEDIA

We're mammals: our young drink milk. All human infants and children are able to do this because their intestines produce an enzyme called lactase. Lactase splits the complex and indigestible *lactose* sugar found in milk into the simpler, digestible sugars *glucose* and *galactose*. For most of the world's population, lactase production shuts down before adulthood.

However, a minority of us—especially in the Americas, Europe, and the Middle East—can continue to drink milk as adults; we're lactose tolerant, having inherited a genetic mutation that maintains lactase production into adulthood. What's interesting is that the groups with the mutation are generally ones descended from populations whose culture included the herding of milk-producing animals, including Europeans, East Africans, and Saudi Arabians. Overall, the adult lactase gene is present in 80 percent of Europeans and 20 percent of Africans and Asians today.

Curiously, it isn't one universal mutation but one of several, all of which result in lactose being produced beyond childhood. That is, over the past 10,000 years, mutations of the lactase gene have arisen *independently* in those parts of the world where our ancestors herded milk-producing animals. Quoting from a 2007 article in *Scientific American*, "Retention [of that trait] in milk-dependent societies illustrates how culture can reinforce the forces of evolution."

So the next time you're grocery shopping, consider this: those dairy shelves are loaded because most of us now living in North America have ancestors who herded cattle, goats, and camels.

36

BARBIE REDUX

That waist-to-hip ratio

"My full-time job is to find me a guy who's going to take care of me," Elizabeth declares, before lending weight to this bracingly feminist cause by saving up for a pair of bigger breasts.

—Anthony Lane, review of *Bad Teacher*, *The New Yorker*, July 4, 2011

So what is it about our—I mean guys'—obsession with women's breasts? Not to mention hips and waists? Have we been brainwashed by our culture to find the "hourglass figure" particularly attractive? Or are we responding to ancient genetic imperatives over which we have little control?

Several complementary studies have pointed to the role of "nature" outperforming "nurture" when it comes to assessing men's gut reaction to the female form. For example, a study of 119 Polish women aged between 24 and 37 found a strong correlation between fertility and relatively low waist-to-hip ratios plus large breasts. During ovulation (when pregnancy is most likely), the level of the female hormone *estradiol* of women with hourglass figures was 37 percent higher on average than women in other groups, estradiol being a good indicator of fertility. According to lead researcher Grazyna Jasienska, "If there are 30 percent higher levels, it means they are roughly three times more likely to get pregnant."

Low waist-to-hip ratio, as presented by two Sigiriya maidens. A sheltered pocket halfway up the 400-foot-high western wall of Sigiriya (Lion Hill) in central Sri Lanka has acted as a protective "gallery" for these fifth-century AD frescoes. AUTHOR

Another study indicates that our obsession with thin-waisted women is neither a recent nor a solely Western phenomenon. After mining a database of 345,000 British and American works of fiction for romantic references to the waist, the late evolutionary psychologist Devendra Singh found that, without exception, slender waists are eulogized. Singh followed this up by asking experts in the field to review ancient Indian and Chinese erotic poetry. Again without exception, every romantic reference to the waist praised its narrowness. (Singh analyzed written texts rather than paintings because clothing can confuse the data. "In poetry you say it outright: Here is a beautiful woman and look at her waist," he explained.)

If you've been following this topic, you may recall it was Singh who first, in 1993, expounded the notion of a universally attractive 0.7 waist-to-hip ratio. His cross-cultural studies indicated that men around the globe generally rated women with low waist-to-hip ratios as optimal mates while noting that a low ratio correlates to lower risk of disease and greater fertility. More recently, it's been shown that abdominal obesity is associated with increased risk of diabetes and breast and ovarian cancer.

So maybe Barbie is trying to tell us something. "I would be the last person to propagate Barbie," says Jasienska, "but when you think about the hourglass shape, Barbie is sort of the symbol." (Would that be before or after her liposuction?)

For what it's worth, I *loved* the movie.

37

SLEEP, PRAY, SLEEP

Are you a normal insomniac?

What do you call a habit of regularly waking up at 1 a.m. and not getting back to sleep until 3 a.m.? *Sleep disorder* perhaps? Not if you happen to be Roger Ekirch, professor of history at Virginia Tech and author of *At Day's Close: Night in Times Past*. His term for such behavior? *Normal.*

After decades of studying letters, memoirs, and diaries from around 1500 to 1750, Ekirch believes that our bodies' "natural" sleep pattern is this: go to bed at nightfall for your *first sleep*, wake a few hours later for a two-hour *watch* (during which time you might pray, talk, or make love), followed by your *second sleep* until dawn. Evidence for this historical sleep–wake–sleep routine goes back to the Greek and Latin writings of Homer, Livy, and Virgil, while contemporary anthropologists find the pattern in many rural parts of the world. In central Nigeria, for example, where the Tiv villagers—who routinely use the phrases *first sleep* and *second sleep*—"wake and talk with anyone awake in the hut," according to one researcher. And when volunteers at the National Institute of Mental Health (NIMH) in Bethesda, Maryland, were deprived of artificial light for several weeks, their bodies reverted to the same sleep–wake–sleep schedule.

According to Ekirch, what we now call "a good night's sleep," six to eight hours of continuous sleep, is an aberration, our body's response to artificial light. While our ancestors normally went to bed soon after sunset, most of us are up for several more hours, thereby depriving ourselves

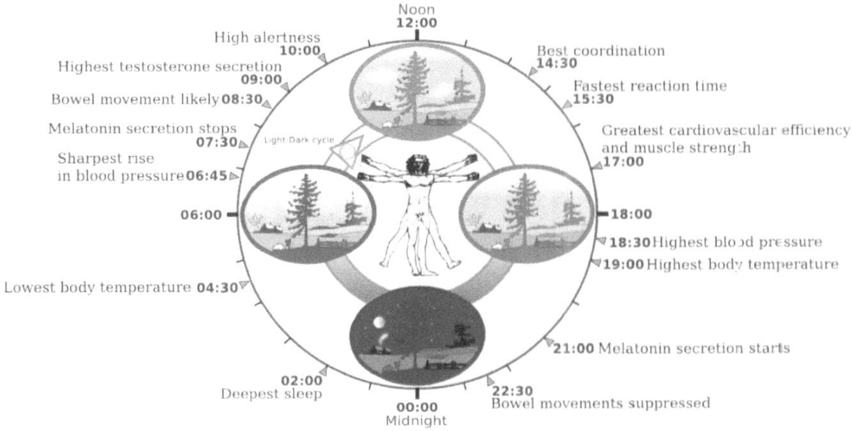

Contemporary human biological clock as modulated by artificial light. YASSINE MRABET, FROM DATA IN *THE BODY CLOCK GUIDE TO BETTER HEALTH* BY SMOENSKY AND LAMBERG, GNU LICENSE.

of the natural middle-of-the-night awake period. According to Harvard chronobiologist Charles Czeisler, "Every time we turn on a light, we are inadvertently taking a drug [melatonin] that affects how we sleep."

Ekirch turned up one particularly tantalizing tidbit in his research: a near-unanimous experience of serenity and contentment during the middle-of-the-night *watch* period. People prayed, meditated on dreams, smoked tobacco, talked, loved. If they could afford candles, they read or wrote. "At no other time during the day or night were distractions so few and privacy so great," writes Ekirch. He quotes Nathaniel Hawthorne: "You have found an intermediate space, where the business of life does not intrude." More prosaically, the NIMH researchers mentioned above found elevated levels of prolactin in volunteers' blood during their 'tweensleep periods. Prolactin is the nurturing, calming hormone responsible for milk production after childbirth, a sort of anti-dopamine. It's what keeps chickens happy brooding on eggs for days at a time.

Most of documented history—the battles, treaties, and betrayals—took place in daytime. Now that we know a little about the "unhistory" of night, we find that our ancestors would have considered our ideal of uninterrupted sleep an abnormal deficiency.

38

WHEN THE OLD WORLD MET THE NEW

Smallpox in America

Imagine this: of 10 friends, nine of them die within a few weeks. Horribly, painfully. You are the only survivor, 1 in 10. That's the extent of the death toll in the Americas, mostly by infectious diseases, starting in 1493. At that time, the population of the New World was probably around 60 million (estimates vary wildly); 150 years later, it was one-tenth of that, 6 million. The extent of the die-off is hard to comprehend, especially because we have so few written records of the devastation.

Columbus's second transatlantic voyage in 1493 was far different from his first. No longer the explorer, he was now a full-fledged slave trader, colonizer, and Christian evangelist. His fleet of 17 ships carried 1,500 men—mostly soldiers—together with livestock: chickens, cattle, pigs, and dogs. Accompanying the humans and animals were invisible pathogens that were deadly to native Americans but to which citizens of Europe, Africa, and Asia had long developed immunity, including typhus, dysentery, malaria, tuberculosis, measles—and, deadliest of all, the viral disease smallpox.

Smallpox, which had killed close to half a billion people in the previous hundred years, was completely eliminated in the 1970s in one of medicine's greatest successes. According to author and explorer Douglas Preston in his bestseller *The Lost City of the Monkey God*, it is "the cruelest disease ever to afflict the human race." Although Preston's particular interest is the abrupt disappearance in the early 1500s of a civilization

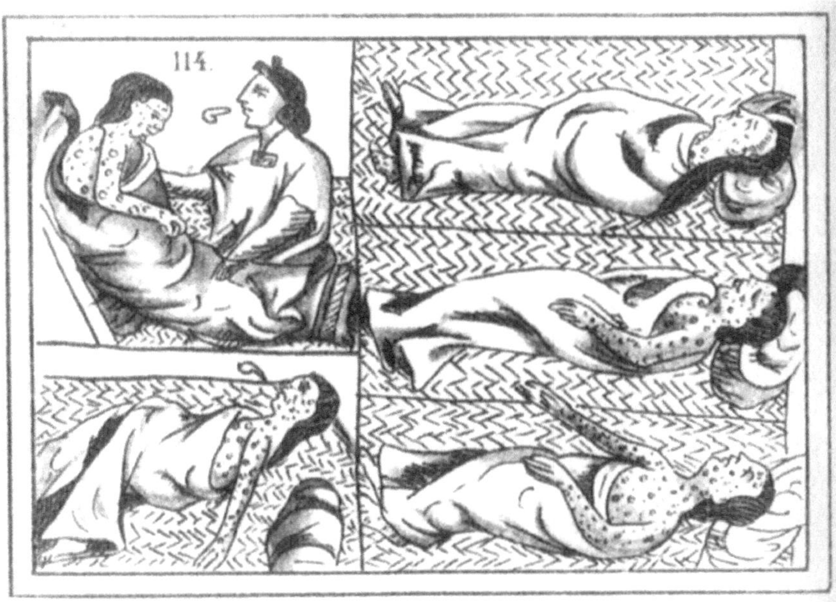

Sixteenth-century Aztec drawing of smallpox victims. PUBLIC DOMAIN

that flourished for centuries in eastern Honduras, it's a much larger story. In the Americas as a whole, where the native population had never been exposed to smallpox, the disease was responsible for more deaths than anything else, including massacres and slavery.

Death from smallpox was excruciating. According to sixteenth-century Seville historian Francisco de Gomara, who interviewed returning New World travelers, smallpox was "a dreadful illness. No one could walk; they could only lie stretched out on their beds . . . when they moved, they screamed in pain. . . . Many died from it, but many died only of hunger . . . for they had no one left to care for them." No wonder Hernán Cortés could defeat the 6-million-strong Aztec Empire with 500 soldiers. As one of Cortés' companions, Francisco de Aguilar, wrote approvingly, "When the Christians were exhausted from war, God saw fit to send the Indians smallpox." Aguilar later became a Dominican monk.

So how was it that Old World residents—Asians, Africans, and European—had immunity while people living in the Americas didn't? In his bestseller *Guns, Germs and Steel*, anthropologist Jared Diamond suggests the one-word answer: *livestock*. In the Old World, domesticated livestock, such as cattle, pigs, chickens, sheep, and goats, were central to virtually every household, meaning that people lived right next to pathogen-carrying animals. When these pathogens mutated to human hosts (for example, cowpox to smallpox or rinderpest to measles), the death toll would have been horrific. However, some were immune, passing their immunity on to their children, resulting in a mostly immune population.

In the Americas, with fewer domesticated animals (mainly llamas and guinea pigs) and more space (New World cities were generally smaller than Old World cities), civilization evolved without coming into contact with these diseases. So when they did, mostly between 1493 and 1640, some 90 percent of the population died. In Preston's words, "The New World was like a vast, tinder-dry forest waiting to burn—and Columbus brought the fire."

39

A BRIEF HISTORY OF VACCINATION

Anti-vaxxers aren't new

Every year, thousands undergo this operation, and the French Ambassador says pleasantly, that they take the small-pox here by way of diversion, as they take the waters in other countries.

—1717 LETTER FROM MARY WORTLEY MONTAGU, WIFE OF THE BRITISH AMBASSADOR TO TURKEY

Lady Montagu, commenting from Adrianople (modern Erdine) in western Turkey, wrote about the annual practice there of *variolation*, the inoculation of healthy children and adults with a weak strain of smallpox in order to confer future immunity from the real thing. Having barely survived smallpox two years earlier while her brother died of it, Montagu became an enthusiastic advocate of the practice on her return to Britain. Over the next 30 years, some 400,000 people in Britain were "variolated," leading to a marked decrease in smallpox deaths. Variolation, which the Chinese had already been practicing for hundreds of years, was the first attempt to induce immunity *before* a disease struck.

Variolation led to vaccination. Unlike variolation, which introduces live smallpox viruses into a body (with a 1 to 3 percent risk of death), vaccination uses weakened or killed germs that cannot cause serious illness. For example, in vaccination's first and greatest success story, smallpox was countered with cowpox, a weak form of smallpox that is nonfatal to humans. The well-known story of how an English country doctor, Edward Jenner, discovered vaccination is worth repeating here.

As Jenner (who wasn't above a touch of self-mythologizing) told it, he heard a milkmaid boast that her face would never be pockmarked by smallpox because she'd already had cowpox. Jenner hypothesized that the girl's cowpox had given her immunity to smallpox, so, in 1796, he deliberately infected the eight-year-old son of his gardener with cowpox using lymph fluid from a cowpox sufferer. Six weeks later, he tried to infect the kid with smallpox (no medical ethics boards back then). The boy remained healthy, and Jenner successfully repeated the experiment on a dozen more children, hence vaccination (from Latin *vacca*, "cow").

Nearly 200 years later, in 1980, the World Health Organization announced that smallpox had been totally eradicated. It's hard to

1802 satirical cartoon from Britain's Anti-Vaccine Society, showing cows emerging from people's bodies after they had been vaccinated with Edward Jenner's cowpox vaccine. JAMES GILLRAY, PUBLIC DOMAIN

exaggerate this accomplishment. In Jenner's day, an average of 400,000 Europeans died—horribly and painfully—from the disease every year. When epidemics struck (which they did regularly), up to one-third of the populations of entire countries died while survivors were pock-marked for life.

Modern vaccines now target a host of other potentially crippling and fatal diseases, including polio, measles, rubella, diphtheria and tetanus, hepatitis A and B, chickenpox, and mumps. Consider that in the pre-vaccine era, almost every family routinely lost a child to one of these diseases; today, childhood death from disease is so rare that it's now thought of as an unexpected tragedy. Following Jenner's discovery, many governments implemented compulsory vaccination programs to protect not only individuals but whole populations. With sufficient (85 to 95 percent) people being inoculated, disease-causing viruses are denied hosts, essentially stopping epidemics in their tracks. (The unfortunate term for this is "herd immunity.")

"Anti-vaxxers" in the nineteenth and twentieth centuries objected on much the same grounds—libertarian, religious, philosophical, and suspicious of science—as they do today. In the United States, after decades of resistance in the late 1800s, the Supreme Court ruled in favor of compulsory vaccination, citing "manifold restraints to which every person is necessarily subject for the common good." That was in 1905, and while it's currently under attack, the law in general puts public health considerations ahead of personal liberty. To put it bluntly, when the exercise of your rights puts other people at risk, you lose those rights.

However, with the anti-vax movement gaining strength in the United States, several diseases that had been completely eliminated there are now making a comeback, including that resulting from the easily transmitted airborne measles virus. Each year now sees hundreds of cases of measles (2,281 in 2025 alone), almost all spread by under-vaccinated individuals.

40

THE ROOTS OF HOMEOPATHY

It's BS (but with the best of intentions)

Sometimes, topics for the weekly science column I write fall, unbidden, into my lap. Such was the case when my wife and I spent a month in Transylvania, in Romania. We'd booked a room in the heart of old Sibiu on Strada Samuel Brukenthal. The name rang a bell for me not because of Brukenthal's 13-year stint as governor of Transylvania thanks to his being a favorite of Empress Maria Theresa (that was news to me) but because he was the patron and mentor of the doctor who founded the popular alternative medicine system of homeopathy.

Samuel Hahnemann (1755–1843) was desperately poor, trying to finish his medical studies in Saxony, in modern-day Germany, when he was taken under Brukenthal's wing. Having passed his exams, Hahnemann moved to Sibiu, where he spent the next two years both as Brukenthal's personal doctor and as cataloger of the governor's vast—280,000 volumes!—library of medieval books. In this latter role, he was exposed to the alchemical lore of such medieval physicians as Paracelus and, notably, Jean Pharamond Rhumelius (1600–1660). Rhumelius was probably the first to propose the homeopathic principle of *Similia similibus curentur*: like cures like.

Hahnemann formalized this aphorism: "That which can produce a set of symptoms in a healthy individual can treat a sick individual who is manifesting a similar set of symptoms." Experimenting on himself with the Peruvian tree bark *cinchona*, he noted that cinchona both causes

Collection of homeopathic substances in Samuel Hahnemann's apothecary, now a pharmacy museum in Sibiu, Romania. AUTHOR

malarial-like symptoms and cures malaria, thus, he thought, proving his hypothesis. (He apparently missed the more mundane explanation, that cinchona contains quinine.)

And so was born the quack science of homeopathy, the general idea being to administer a super-dilute, very specific, "homeopathic" remedy for an illness. Yet Hahnemann was far from a quack doctor. Rather, he was one of the brightest and most conscientious medical researchers in late eighteenth-century Europe. Master of many languages (he regularly supported himself as a translator) and a prolific writer, he was appalled at what then passed for medical science. "After I had discovered the weakness and errors of my teachers and books," he wrote, "I sank into a state of sorrowful indignation, which had nearly altogether disgusted me with the study of medicine."

At its heart, homeopathy, along with other "energy field" practices such as chiropractic, Reiki, chakra healing, and therapeutic touch, represents the last stronghold of the ancient theory of vitalism. This is the belief that living organisms contain some nonmaterial energy, often referred to as *élan vital* (or, perhaps, soul) absent in inanimate entities. So the fact that a homeopathic solution is diluted to the point where no molecules of the original substance remain is of no concern. In fact, for Hahnemann and his modern followers, that's the whole point: the dilution removes the harmful effects of a substance while, in his words, "arousing the spirit-like medicinal powers of the crude substances."

Two hundred years later, vitalism is all but dead, and materialism (that is, "We're just rearranged food!") reigns. Homeopathy, according to mainstream medicine, is, at best, a placebo. And Hahnemann, insofar as he was dissatisfied with the medical science of his time and tried to fix it? He was a hero.

41

AN UPLIFTING STORY

Is the bra a booby trap?

A rite of passage for a girl is, I understand, putting on her first bra. The equivalent life event for a boy is, of course, the removal thereof, once he's figured out how to unsnap the hook-and-eye catch, one-handed and in the dark. Somehow most of us survived these early experiments, and the fact that most women wear bras is probably as unremarkable as brushing our teeth.

Bras have been around at least since ancient Roman times in the form of wool or linen breast bands (Latin *mastodesmos*). Recently, two modern-looking bras were found in a stash of medieval garments in an old castle in Austria. Dated to 1450, they're similar to "longline" (corset-style) bras, with shoulder straps and eyelets for fastening at the back with string. Etymologically, bra, short for *brassière*, was once French for an underbodice, coming from *bracière*, Old French for a military arm protector. (*Bras* means "arm" in French.) This evolved to a soldier's breastplate and later a woman's corset.

Despite the ubiquity of bras today, several concerns have been raised about them over the past 25-odd years:

- Back pain: One hundred women who complained of back pain were invited to stop wearing their bras for two weeks, according to a five-year study published in 2000 in the professional journal *Clinical*

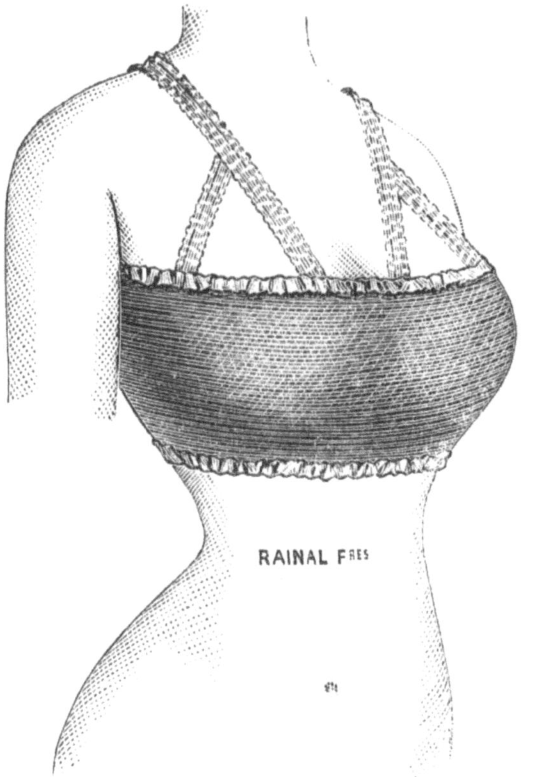

1907 French elastic *brassière* "for people who can dispense with the use of the corset. It is also used for the night, for seaside bathing or in cases of extreme breast tenderness." WIKIMEDIA COMMONS

Study of Pain. With no weight on their shoulders, most participants experienced pain relief, and three years later, 79 percent of them had stopped wearing a bra "to remove breast weight from the shoulder permanently because it rendered them symptom free."

- Incorrect fit: Researchers checked the bra sizes of large-breasted women requesting mammoplasty (breast reduction) in an investigation published in 2002 in the *British Journal of Plastic Surgery*. Of the 102 women in the study, *all* were wearing incorrect bra sizes. On average, the cups were three sizes too large and the bands four inches too small. According to the researchers, "Obesity, breast hypertrophy [extreme enlargement], fashion and bra-fitting practices combine to make those women who most need supportive bras the least likely to get accurately fitted bras." Additionally, bras come with matching cups despite some 25 percent of women having breasts that differ by one or more cup sizes, the left breast usually being the larger.

- Breast sag: Several studies conducted at the Centre Hospitalier Universitaire, Besançon, France, have shown that bras actually contribute to breast sagging rather than alleviating it. For example, Professor Jean-Denis Rouillon followed the anatomy of 330 women over 15 years. Some wore bras, some didn't. He concluded that bras cause the chest muscles supporting the breast to partially atrophy. "Medically, physiologically, anatomically—breasts gain no benefit from being denied gravity," he said in an interview in the British *Telegraph* newspaper. The breasts of women who normally went braless sagged three inches less, on average, than those wearing bras.
- Natural development: Another long-term Swedish study led by Professor Göran Samsioe of Lund University determined that bras hampered the natural development of breast-supporting elastic tissue, stating, "If natural movement is restricted by a bra that is too tight, it can affect the growth of these tissues . . . there is no medical reason for [bras] and in some cases it can even be harmful."

Given our culture's phobia about seeing breasts (and nipples) move naturally under clothing, these findings are unlikely to lead to a sea change in bra wearing. One can hope, however, that smart medicine will one day outvote the dictates of fashion and modesty.

INVENTIONS

Building better mousetraps

42

THE REAL WHEEL

Someone had to be first

Imagine that a time machine has transported you back 5,000 years to ancient Egypt and you're watching a crew of workers wrestling with a big block of stone. They're building a tomb for some A-list dude—before pyramids, there were lower, simpler *mastabas* (flat stone platforms)—and you notice their system for transporting the block from quarry to tomb: log roll. As the heavy block rolls off the rearmost log, a couple of guys pick it up and place it at the front so that the stone is always supported on three or four rollers. You say to them, "Wouldn't it be a whole lot easier if you replaced the logs with a flat cart on wheels?" "What's a wheel?" they ask.

While the step from logs (cylinders) to wheels (discs) looks pretty obvious to us, it was a technological stroke of genius, one so big that the wheel may have been invented only once before the knowledge spread throughout Eurasia and the Middle East. The birthplace of the wheel is disputed. Linguistically, the first use of the Proto-Indo-European word for *wheel* comes either from present-day Ukraine or from Turkey, depending on who's making the case. Archaeologically, the earliest image of a wheel was found in Poland, while the oldest actual wheel is from Mesopotamia, dating to around 3500 BC.

The leap from roller to wheel had to wait for the invention of metal tools needed to craft holes in wheels and to shape the ends of axles. In

Onager-drawn cart on the Sumerian "Battle Standard of Ur," about 2500 BC.
PUBLIC DOMAIN

order for wheels to turn smoothly, the holes and axle ends had to be perfectly round and smooth, snug enough to keep the wheel in place but not so snug as to prevent it from turning. The earliest metal tools were cast from copper, with copper chisels making their first appearance in the archaeological record around 4000 BC, while tougher bronze (approximately 90 percent copper and 10 percent tin) had to wait for another 1,000 years.

That wheel found in Mesopotamia was probably a potter's wheel, which, being horizontal, didn't have to bear as much load as a vertical wheel on a cart. The introduction of such wheels would have been a huge boon to early pottery makers, who could then turn out pots in far greater numbers than previously. Clay pots—rat-proof containers and water carriers—would have made the day-to-day challenge of survival that much easier.

Once our inventive forebears had figured out how to mount a pair of wheels on the end of an axle, carts drawn by onagers (wild asses) and oxen soon followed. The earliest wheels used for transportation were solid slabs of wood from mature trees. The next technological advance came around 2000 BC with the invention of the spoked wheel. Before then, solid wheels facilitated slow transportation; after, the new technology led to nimble horse-drawn chariots, the ancient equivalent of Panzer light tanks.

Records left by both sides, Egyptian and Hittite, following the Battle of Kadesh (1274 BC), attest to the utility of these mobile war machines, in which *6,000* chariots reputedly clashed in present-day Syria on the banks of the Orontes River. The result was a standoff, with the leaders of both sides, Rameses II and Muwatalli II, claiming victory for their own forces.

Finally, fast-forward some 3,000 years. Advances in metal technology led to the steel-spoked tension wheel, ubiquitous on early automobiles and on the most wonderfully efficient vehicles of all: *bicycles.*

43

UNWRAPPING THE PAST

Reading a lump of charcoal

If you happen to be a time traveler, do avoid Italy's Bay of Naples on August 24, AD 79. That's when Mount Vesuvius, which had been dormant for 800 years, blew its top, engulfing several nearby Roman cities in a thick layer of volcanic ash. The catastrophe turned out to be a boon for future archaeologists. The most famous of the buried cities, Pompeii, was rediscovered only in 1599, and today, you can wander its streets, enter into mosaic-rich courtyards, and read graffiti on its walls.

Nearby, less visited Herculaneum (modern Ercolano) is the home of the Villa of the Papyri, so called for some 1,800 papyrus scrolls found there shortly after the 1709 serendipitous discovery of the city by workers digging a well. However, unlike well-preserved papyrus scrolls from sites in Egypt and the Near East, Herculaneum's papyri were burnt to a crisp during the eruption. Over the years, archaeologists have tried, disastrously, to unwrap the carbonized lumps; they crumble at the lightest touch. Now, using really cool techniques (the next best thing to time travel), researchers are bringing the ancient writings back to life.

In order to read a carbonized papyrus without damaging it, the charred black lump that once was a 20-foot-long scroll is scanned noninvasively with X-rays, one superthin slice at a time. The resulting three-dimensional digital model is then "virtually unrolled" to reveal the original writing. The technique involves two modern breakthroughs:

The carbonized En-Gedi scroll as found in 1970 (right) and part of it virtually unwrapped, revealing Hebrew characters from the Book of Leviticus. SEALES ET AL., © UNIVERSITY OF KENTUCKY, USED WITH PERMISSION

microtomography (three-dimensional X-rays, routinely used in hospital CAT scans) and sophisticated software to process the resulting huge database of information.

Computer scientist Brent Seales created the software. He and his team at the University of Kentucky pioneered the virtual unrolling technique on a scroll found in 1970 at the ancient En-Gedi synagogue on the western shore of the Dead Sea. A fire in the sixth century AD destroyed the complex, but in the wreckage, archaeologists found the synagogue's Holy Ark, containing a lump of baked papyrus. Israeli authorities had presciently stored the En-Gedi scroll against the day when scientific progress could allow decipherment of the fragile object. Forty-five years later, Seales's team was able to do just that, scanning and virtually unwrapping the carbonized scroll without damaging it. In 2015, they announced their findings after reading—on a computer screen—a 1,700-year-old Hebrew text from the biblical Book of Leviticus.

The En-Gedi fragment was an easier challenge than the Herculaneum library scrolls because the Jewish scribes used ink containing iron, which stands out since iron absorbs X-rays more than burnt papyrus. In contrast, ink used on the House of Papyri scrolls is carbon based, and burnt papyrus is essentially carbon, so differential absorption couldn't be used. Instead, scientists used a new technique, X-ray phase contrast

Reading burnt scrolls: the black form-fitted case holds a scroll vertically while rotating it through 360 degrees. X-rays enter from behind and to the right of the viewpoint and pass through the scroll, striking the X-ray detector on the opposite side, where the red laser is hitting. The resulting X-ray images are used to reconstruct the interior of the scroll. COURTESY EDUCELAB/UNIVERSITY OF KENTUCKY. Inset: How they tried (mostly disastrously) to unroll the Herculaneum scrolls 200 years ago. PUBLIC DOMAIN tomography, to read the scrolls by measuring the difference in refraction between the ink and the underlying papyrus.

Researchers have just started to analyze the Herculaneum scrolls. So far, they've discovered 2,000-year-old texts written by a follower of Epicurus, the Greek philosopher who promoted *ataraxia*, peace of mind cultivated by living self-sufficiently in the company of friends. Hundreds of scrolls remain to be digitally read, with possibly thousands more still to be found, potential treasure troves from the past, awaiting rediscovery in the unexcavated parts of Herculaneum.

I bet Epicurus would have enjoyed the idea of his words being reread by future generations—perhaps finding *ataraxia* in their contemplation.

44

A BRIEF HISTORY OF DILDOS

A penetrating account

Thou . . . madest to thyself images of men, and didst commit whoredom with them.

—Ezekiel 16:17, KJV

Along with notable human achievements such as the invention of the plow and the wheel, we should also celebrate a much older innovation, by tens of thousands of years: the humble dildo. In 2005, a team of researchers from Tübingen University in Germany unearthed a foot-long, four-inch-circumference, polished siltstone phallus from the Hohle Fels cave near Ulm. Dated to 28,000 years ago, it's now exhibited in the Blaubeuren Museum of Prehistory, labeled "Ice Art—Clearly Male."

If you're bothered by the vision of a woman pleasuring herself with a length of hard cold stone (a far cry from today's silicon devices), you can take heart from this reclassified wooden object from the Roman fort Vindolanda near Hadrian's Wall in northern England. The wall was built around AD 125 by order of Emperor Hadrian to solidify Roman control of most of present-day England and Wales, acknowledging that pacifying the tribes in today's Scotland was impracticable given his limited military resources. (It's not too different today.)

The wooden object was found in a ditch by researchers from the University of Newcastle. They originally—and sweetly—classified it

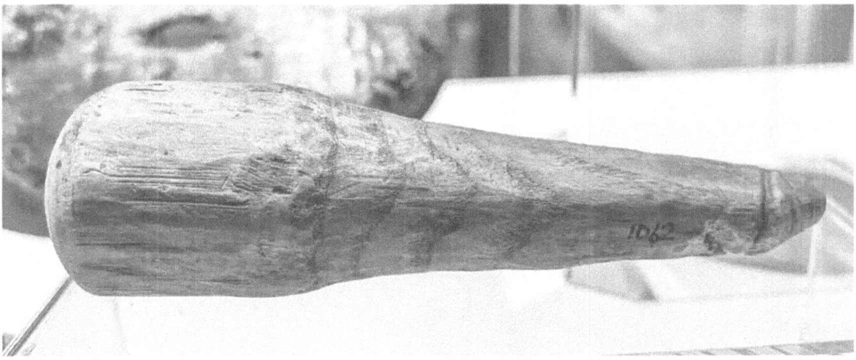

Not a darning tool. THE VINDOLANDA TRUST, USED WITH PERMISSION

as "a darning tool." Newcastle University archaeology senior lecturer Rob Collins was baffled by this, saying, "It's kind of self-evident that it is a penis. I don't know who entered it into the catalogue. Maybe it was somebody uncomfortable with it." It seems to be unique, being the only known life-size Roman dildo ever found. It's just six inches long, although researchers say it was originally larger, having shrunk with age. (This writer can empathize.) The ends are very smooth, "indicating it was used for something over a period of time," per the coy press release.

As an aside, size does matter though perhaps not in the way male readers might expect. A 2015 study employed dildos of many sizes to learn the average size that a sample of 75 women preferred for their partner's penis. Results: 6.3 inches long, 4.8-inch girth, just slightly larger than average.

Although no other Roman dildos have been found, they are frequently depicted in Roman art, particularly in Pompeii, the city in southern Italy buried under volcanic ash in AD 79. There, graphic representations of dildos (and much else) abound, leaving no doubt as to how they were to be used. Similarly with the ancient Greeks. More than 400 years earlier, Athenian audiences would have been tickled pink, so to speak, by Aristophanes' comedy *Lysistrata*. The titular heroine persuades the sisterhood to deny sex to their menfolk until Athens and Sparta make peace: "Meanwhile we can use our eight-fingered leather dildos. At least they'd serve as a sort of flesh-replacement for our poor cunts."

Dildos have come a long way from the Upper Paleolithic to today's do-it-yourself three-dimensionally printed models. I'll leave you with one bit of trivia to illustrate how far the United States, Land of the Free, hasn't come with respect to equal rights for women. In 1998, the State of Alabama passed a law banning the distribution of "any device designed or marketed as useful primarily for the stimulation of human genital organs." Nine years later, the U.S. Supreme Court left the law intact.

45

PERPETUAL MOTION MACHINES

Fighting thermodynamics

Around 1180, Indian mathematician Bhaskara II invented what has been claimed to be the world's first perpetual motion machine. It was a vertical wheel with mercury-filled tubes attached to the rim; as the wheel turned, mercury flowed to the lowest parts of the tubes, maintaining the rotation indefinitely—you can see a modern version of its action at https://www.youtube.com/watch?v=7KqOwJKWIAw. The free energy thus generated could solve today's energy crisis! No more fossil fuels polluting the atmosphere and warming the planet! No more messy gas or diesel engines on our roads! Simply scale up Bhaskara's wheel and enjoy a Golden Age of Energy.

If only. I'll leave it to you to figure out how the wheel in the video keeps turning, but even if you don't watch it, you already know it's a scam, right? From Bhaskara on down to the present time, inventors have been presenting the world with perpetual motion machines, that is, machines whose output—in this case a revolving axle that could be used to generate electricity—exceeds its input. No energy goes in, some energy comes out. And there goes the first law of thermodynamics. Remember? Energy can be neither created nor destroyed; that is, there's no free lunch. Perhaps that should be amended: the only free lunches are eaten by inventors who are able to con rubes into investing in their "free energy" machines. (Next up: automobiles that run on water.)

Norman Rockwell's 1920 *Popular Science* cover shows a modern version of Bhaskara's wheel, ca. AD 1180. Can you see why it can't work? Hint: Where is its center of gravity? PUBLIC DOMAIN

A twentieth-century version of Bhaskara's wheel is shown in the "mass leverage" contraption shown here, in which the metal balls on the right have more leverage than those on the left, creating perpetual rotation. Except four balls on the right (in this case) are counterbalanced by five on the left. More to the point—and common to all "wheel" types of perpetual motion machines—the center of gravity of the moving part is lower than the central axle, meaning that, once started, it will end up swinging to and fro, like a pendulum, before friction causes it to stop.

But wait, I've been talking about perpetual motion *machines*. What about just plain perpetual motion, that is, something that keeps moving eternally without doing any actual work? Isn't that possible? Afraid not. Now we're up against the second law of thermodynamics, which says—for our purposes—that a closed system can never run at 100 percent efficiency no matter what; that is, you can't even break even.

Meaning that nothing works forever, whether it's a pendulum (losing energy through friction of its pivot and of air molecules), a "drinking bird" toy (which runs until all the water is evaporated), Earth spinning in space (whose kinetic energy is constantly being lost to tidal friction, hence the insertion of about 30 "leap seconds" since 1972), or a hydrogen atom (whose single proton, with a half-life of 10^{-33} years, will—eventually!—decay).

None of this is going to stop inventors trying to outwit the laws of thermodynamics—look no further than YouTube if you don't believe me. Little has changed since Leonardo da Vinci bemusedly wrote, more than 500 years ago, "Oh ye seekers after perpetual motion, how many vain chimeras have you pursued? Go and take your place with the alchemists."

46

SPECTACLES AND TELESCOPES

A small mystery

Bird-watchers—twitchers, to their friends—are immediately recognizable in the great outdoors by having fancy binoculars strung around their necks plus, often, a digital single-lens reflex camera sporting an outsize telephoto lens. Both devices employ essentially the same system of optics to make distant objects appear closer, a system that dates back about 400 years.

Most reference books give the date for the invention of the telescope as 1608, and it's usually attributed to Hans Lipperhey of Middleburg in the province of Zeeland, Holland. Two years later, a hitherto unknown professor of mathematics in Padua, Italy, published a book outlining the discoveries he'd made while observing the night sky through his homemade telescope. Galileo Galilei became an instant celeb, his book *Siderius Nuncius* (*The Starry Messenger*) was a bestseller, and the modern science of astronomy was born.

The mystery is this: why was there a gap of more than 300 years between the invention of spectacles and the invention of the telescope? After all, a telescope, in its simplest form as used (and improved) by Galileo, is simply two spectacle lenses, one to correct near vision and one to correct far vision, held a few inches apart. You can easily make a telescope for yourself with a pair of distance (concave) glasses and a pair of reading (convex) glasses. Hold the reading glasses at arm's length (your objective

The earliest known representation of spectacles is found in the chapter house of San Nicolo, Treviso, Italy. This 1352 fresco by Tomaso da Modena shows the Dominican monk Hugues de St. Cher copying a manuscript. © THE COLLEGE OF OPTOMETRISTS/NEIL HANDLEY, USED WITH PERMISSION

lens) while looking at them through the distance glasses (your eyepiece). Congratulations—you've just created a telescope!

We know that convex spectacles to compensate for presbyopia (farsightedness) were invented at the end of the thirteenth century. Presbyopia affects most of us at some point in our lives, as the cilia muscles of our eyes weaken to where we can no longer focus on something close, such as a book. The invention of spectacles for reading must have been an incredible boon back then, especially for aging monks engaged in the painstaking work of hand copying manuscripts before the invention of printing. Concave spectacles, to correct myopia (nearsightedness), were invented a little later.

Researchers believe that two factors were responsible for the long gap between spectacles and the telescope: the poor quality of the glass then available and the inaccuracy of the finished lenses after grinding. In fact, some historians of science believe that the *idea* of the telescope, in the form of two lenses—concave and convex—was known decades before 1608 but that it was only then, in Holland, that the state of the art of glassmaking allowed the creation of a *useful* instrument.

As with all inventions, theory had to be accompanied by practical technology to turn a great idea into a useful device.

47

GUTENBERG'S LEGACY

Inventor or synthesizer?

Although Johannes Gutenberg (1398–1468) is usually credited with having invented printing, it's fairer to say that he introduced a *system* of mechanical printing to Europe since virtually all the technologies he employed had already been invented elsewhere in Europe or China. Like most innovators, his flair lay in combining existing methods, modifying and improving as he did so. The revolution in printing instigated by Gutenberg could have happened a century earlier had someone of his genius lived then, or it could have happened later since he was one of about half a dozen Europeans actively researching ways to mechanize printing in the mid-1400s, and surely one of them would have figured out a system if Gutenberg hadn't beaten them to it.

The success of an innovation can be gauged by the speed with which it's adopted. Within 50 years of the publication of Gutenberg's 1,200-page, 42-line Bible in 1455, at least 250 printers were furiously at work all across Europe using his technology. Historians have estimated that 15 million books were printed mechanically between 1455 and 1500, more than the sum total of all books painstakingly copied by hand before then.

The revolution in printing and, with it, mass communication extended to virtually every facet of life. It's hard to imagine the Age of Enlightenment or the Scientific Revolution without the ready means to exchange information. Prior to Gutenberg, literacy was considered mysterious or even supernatural by the vast majority of people who

The first copy of the Gutenberg Bible brought to the United States, now in the New York Public Library. PUBLIC DOMAIN

couldn't read or write. After him, an unstoppable cycle came into play: books—lots of books, cheap books—encouraged literacy, while literacy in turn stimulated the demand for books.

CHANGING OUR MINDS

The ease with which newly literate people could communicate revolutionized how minds connected to minds. In his book on the internet revolution, *The Shallows: What the Internet Is Doing to Our Brains*, technology writer Nicholas Carr says, "As books became common, men could look more directly at each other's observations, with a great increase in the accuracy and content of the information conveyed."

Not just factual information, either, but fiction too. Think about the last time you lost yourself in a novel as you essentially re-created the author's thoughts in your own brain. That intimate mind-to-mind connection at a distance was all but impossible before Gutenberg. After the printing revolution and the rapid democratization of literacy that followed, authors found ways to share their innermost feelings by writing with "a heightened awareness and closer observation of sensory experience that was passed on in turn to the reader," per historian Elizabeth Insentient.

Carr claims that the proliferation of books that began with Gutenberg actually changed how we think—for the better, in his opinion. More recently, however, he believes that the internet is modifying brains, this time for the worse. Increasingly, we're reading in bite-sized chunks, skimming rather than absorbing, readily distracted by sidebars, pull quotes, inset videos, and ads. And not just when we're online; today, magazines and newspapers are following the trend of computer screens in catering to our novelty-seeking minds. (Ironically, Carr's book is available on Kindle for $9.41.)

FIVE-PART SYSTEM

As I said, Gutenberg's genius was to combine several existing technologies, adapting and improving them in the process, thus creating a printing system. The five most important ones are the following:

- Printing press: Printing involves transferring ink from type to paper. Gutenberg adapted the traditional screw-driven wooden wine presses of his time to a simple downward motion. This avoided the smudging that would have otherwise resulted from the screw rotating as the type pressed against the paper or vellum.
- Metal alloy for type: Before Gutenberg, type was composed of bronze or tin. Gutenberg, who earlier had been a goldsmith by trade, discovered (or learned from someone else) that the addition of antimony to a mixture of lead and tin minimized shrinkage and cracking as the molten type pieces cooled.
- Hand-cast molds: Of all Gutenberg's inventions and adaptations, his design for a flexible, reusable hand mold for casting type was probably the most important. His molds could be used multiple times to cast identical copies of the same letter rapidly and economically.
- Movable metal type: Printing on wooden blocks was invented around AD 200, while the British Museum's beautiful copy of the Buddhist Diamond Sutra, the oldest dated printed book, is from AD 868. Sometime around AD 1100, movable type—individual

symbols on separate blocks—was invented in China using wood and ceramic cubes. Later, in 1377, more durable movable metal type was used in Korea to print the *Jikji* book of Buddhist teachings. However, the thousands of individual characters needed hindered progress in mechanical printing in the East. Gutenberg and his successors in Europe, however, needed fewer than 80 letters, numbers, and punctuation marks.

- Oil-based ink: Traditionally, scribes used water-based inks to copy manuscripts, which also worked for woodblock printing. Water-based ink doesn't adhere to metal, however, so Gutenberg adapted the oil paints used by artists to formulate ink from a combination of lampblack, linseed oil, and egg white.

1568 woodcut of early wooden printing press. PUBLIC DOMAIN

By taking these separate elements and combining them into an integrated system, Gutenberg was able to produce books accurately, rapidly, and cheaply compared with manual copying.

For the first few decades of mechanical printing, books looked and felt much the same as they had before Gutenberg: big and heavy with a fat Gothic typeface to match. The printer and humanist Aldus Manutius changed that in 1501 with his pocket-size copies of Greek and Roman classics, using two innovations: a light italic typeface that increased both legibility and the number of words on a page; and folding the printed sheets into eight parts, cutting them after printing to produce the "octavo" size. Other than Manutius's breakthroughs, Gutenberg's mechanized system of book production remained essentially unchanged for 300 years until the 1800s, when steam power ushered in a secondary revolution in printing led by Friedrich Koenig.

Apparently, most of us still prefer reading tree-based books over computerized pixels on a screen, but the era of Everyman curling up in a chair with a delicious novel, entering the author's heart and soul, may be coming to an end. Gutenberg's gift may have been a temporary one, lasting about 500 years, as our easily distractible and malleable brains adapt to the new paradigms of our digital age.

48

ATOMS: SOLVING THE INVISIBLE

Inventing our theory of matter

The story of how science came to understand what we—and everything—are made of effectively began with the "father of chemistry," French nobleman Antoine-Laurent Lavoisier. Before losing his head to the guillotine during the French Revolution, Lavoisier recognized that stuff came in two varieties: that which could be broken down into simpler stuff (*compounds*) and that which couldn't (*elements*, as in "elementary"). This simple yet momentous observation prepared the groundwork for the science of chemistry.

Ten years later, in 1803, English gentleman-scientist John Dalton quantified Lavoisier's insight. Reviving the Ancient Greek term for indivisible atoms (*a-tom*, "that which can't be cut"), Dalton's genius was to suggest that what distinguishes the properties of one element from another depends on the mass of each of its constituent atoms and that compounds are formed by a combination of elements in simple whole-number ratios. For example, two atoms of hydrogen combine with one atom of oxygen to make a water molecule.

THE PERIODIC TABLE

We now know (1) that an atom is composed of a cloud of light, negatively charged electrons surrounding a heavy, dense nucleus consisting of positively charged protons and neutral neutrons and (2) that one element

1 H																	2 He
3 Li	4 Be											5 B	6 C	7 N	8 O	9 F	10 Ne
11 Na	12 Mg											13 Al	14 Si	15 P	16 S	17 Cl	18 Ar
19 K	20 Ca	21 Sc	22 Ti	23 V	24 Cr	25 Mn	26 Fe	27 Co	28 Ni	29 Cu	30 Zn	31 Ga	32 Ge	33 As	34 Se	35 Br	36 Kr
37 Rb	38 Sr	39 Y	40 Zr	41 Nb	42 Mo	43 Tc	44 Ru	45 Rh	46 Pd	47 Ag	48 Cd	49 In	50 Sn	51 Sb	52 Te	53 I	54 Xe
55 Cs	56 Ba	57 La	72 Hf	73 Ta	74 W	75 Re	76 Os	77 Ir	78 Pt	79 Au	80 Hg	81 Tl	82 Pb	83 Bi	84 Po	85 At	86 Rn
87 Fr	88 Ra	89 Ac	104 Rf	105 Db	106 Sg	107 Bh	108 Hs	109 Mt	110 Ds	111 Rg	112 Cn	113 Uut	114 Uuq	115 Uup	116 Uuh	117 Uus	118 Uuo

58 Ce	59 Pr	60 Nd	61 Pm	62 Sm	63 Eu	64 Gd	65 Tb	66 Dy	67 Ho	68 Er	69 Tm	70 Yb	71 Lu
90 Th	91 Pa	92 U	93 Np	94 Pu	95 Am	96 Cm	97 Bk	98 Cf	99 Es	100 Fm	101 Md	102 No	103 Lr

Periodic table of the elements. COURTESY PRINTABLEPALS.COM

is distinguished from another principally by the number of protons in its nucleus. Remember the periodic table of the elements on the wall of your school science lab? Dalton's intellectual successor, Russian chemist Dmitri Mendeleev, came up with this ingenious way of arranging elements according to weight (left to right) and properties (top to bottom). Consider three examples:

- The lightest element, *hydrogen*, atomic number 1 (whose nucleus contains just one proton), is upper left.
- All organic compounds contain *carbon*, atomic number 6 (six protons in each nucleus).
- *Uranium*, atomic number 92 (92 protons), is the heaviest naturally occurring element.

The beauty of the periodic table is its universality. Everything (everything!) you touch, see, smell, and taste consists of either pure elements or, more likely, a mixture of the elements found in this deceptively simple matrix. Take your body, which is made up of about 65 percent oxygen, 18 percent carbon, 10 percent hydrogen, and 3 percent nitrogen by mass, while the remaining 4 percent consists of traces of almost every other naturally occurring element in the table.

Since the time of Lavoisier, Dalton, and Mendeleev, chemists, physicists, and astronomers have made huge strides in understanding the origin of elements. Nearly 14 billion years ago, soon after the creation event we misleadingly term the Big Bang (neither big nor a bang), the universe consisted only of the lightest elements hydrogen, helium, and lithium (one, two, and three protons, respectively). No carbon, no oxygen, no zinc, no iron, no uranium; these would be synthesized from the original three elements. *How?* By nuclear fusion. *Where?* Inside stars.

STAR STUFF

Many technical problems needed to be solved to arrive at a complete understanding, so detailed answers have been available only in the past 60-odd years. The biggest hurdle physicists had to overcome is right here in your blood: iron.

Although it may seem as if the main business of stars is to pump out light, their *real* work is to synthesize heavy elements from lighter ones. In a process that diet-pill manufacturers might dream about, our sun is constantly losing mass. The intense temperature and pressure in the heart of our star forces nuclei of the lightest element, hydrogen, to combine, or fuse, into nuclei of the next heavier element, helium. Every second, the process releases 5 million tons of mass. What we perceive as light and heat is the result of this "surplus" mass being converted into energy, Einstein's $E = mc^2$.

As the sun and other medium-sized stars like ours reach old age, a secondary process transforms the helium "ash" produced during the stars' prime years into carbon, which is heavier yet. And as far as our sun is concerned, that's it. Carbon is the heaviest element that can be synthesized in stars as massive as the sun. To produce heavier elements, you need stars with masses greater than about eight times that of the sun since more mass equates to higher temperatures and pressures in a star's core, resulting in the fusion of ever-heavier elements. In massive stars, the ash from each phase of the process provides fuel for each successive phase until you get to the most stable element (because its atoms are most tightly bound): iron. So at first blush, it might seem that iron is the end of the line. Until you get to iron, the fusion process *releases* energy, but to synthesize elements heavier than iron, you need to *import* energy. Which is where supernovae come in.

At the end of its life, every star becomes unstable. While it lives, any star—our sun, for example—is in a state of equilibrium: gravity holds it all together and creates intense pressure in the core (hence fusion), while gas pressure from the fusion prevents the star from collapsing. But once iron is synthesized in a star having a mass greater than eight times that of the sun, the game's as good as over. The iron core grows like a cancer in the heart of the star, damping nuclear reactions in all that it touches. The star becomes fatally imbalanced and falls in on itself, collapsing into an incredibly dense "neutron star."

As the outer layers fall in toward the dense core, that core bounces back from its initial collapse. The result is cataclysmic, a fury of shock waves as the inward-rushing material meets the rebounding core. The energy released is so intense that to us it seems as if a new star—sometimes as bright as the entire parent galaxy to which it belongs—suddenly appears in the sky: a supernova. In all this explosive chaos, vast quantities of heavier-than-iron elements are produced. While it took billions of years to synthesize the light elements, the heavy ones are created in minutes. And in the aftermath of a supernova, new stars form, phoenix-like, from the dross of the old.

Nearly 5 billion years ago, a star we call the sun coalesced out of dust and gas consisting mostly of the remnants of one or more previous generations of supernovae. Material destined to become planets and moons swirled around this protosun, a cloud consisting of basic hydrogen, helium, and lithium and heavier star-formed stuff. Eventually, on at least one of the planets, life began.

Billions of years later, an exaltation of those atoms, an organism with iron in its blood, calcium in its bones, and iodine in its thyroid, is reading this account.

49

MONSIEUR SAX'S BRASS CLARINET

Creating that smooth sound

Adolphe Sax, were he alive today at 200 and something years old, probably wouldn't be surprised that many rappers feature the saxophone on their tracks—Macklemore's 2012 hip-hop hit "Thrift Shop" is a prime example—because he designed it for versatility. By combining the agility of woodwind with the projection of brass, Sax created an instrument that is at home in virtually every musical genre. With a single reed mouthpiece, the saxophone is officially a woodwind instrument even though it's made of brass. In fact, in its soprano form, the sax looks like a brass clarinet.

The soprano sax is what drew me to the instrument in the first place, my 10-year-old self transfixed by Sidney Bechet's "Summertime" on my parents' wind-up gramophone in all the high fidelity (!) of a 10-inch shellac "78" record. My early flirtation with jazz was soon swept away by Bill Haley, Elvis, Fats Domino, Jerry Lee Lewis. Rock 'n' roll was my thing in those awesome (and awful) pubescent years, and rock bands had plenty of use for the sax, with most combos including one or two of them to help pump out that insistent 4/4 beat. Even if you didn't have the first clue about dancing, you just had to get up and jive.

Somewhere in late 1950s, though, Paul Desmond wrote "Take Five" for the alto sax, which became the biggest-selling jazz single of all time. I loved it, at the same time being vaguely aware that someone called John Coltrane played a mean tenor sax on his "Giant Steps" LP (remember LPs?). The sax kept popping up in popular culture too. I hear the phrase "Pink Panther," and Henry Mancini's chromatic Pink Panther sax theme from 1963 plays in my head. A year later, Stan Getz's "Girl from Ipanema" again showed the sax's versatility as it embraced the bossa nova. The 1970s brought us Bobby Key's tenor sax solos on such Stones' hits as "Brown Sugar," Bob Seger's opening alto wail on "Turn the Page," and Clarence Clemons' counterfoil to the Boss's guitar on Springsteen's *Born to Run* album.

Saxophones are found just about everywhere there's music: in the classical repertoire (Ravel's *Bolero* and Mussorgsky's *Pictures at an Exhibition*), in military bands (many of Sousa's marches), in William Lovelock's chamber music, in bebop (pioneered in 1939 by jazz sax pioneer Coleman Hawkins in *Body and Soul*), in Lester Young's "cool" jazz, and of course in rock, pop, and rap.

Structurally, the sax is a thin brass conical tube, flared at the end, with about 20 tone holes covered by keys that the player opens or closes (some holes are open by default, some closed). In the alto, tenor, and baritone versions, the tube is bent into the traditional folded U, while the soprano sax is usually straight.

Adolphe Sax, from Dinant and later Brussels, Belgium, had been looking to fill the gap between woodwind and brass instruments. Before inventing the saxophone, he was already experimenting with a precursor of the tuba (the last brass instrument to join the orchestra) and the euphonium (the bass "saxhorn"). Both his parents were instrument designers whose inventiveness resulted in the contemporary French horn.

Adolphe learned to play the clarinet early on, experimenting with a design for a bass clarinet that would probably have won the gold prize in the Brussels Industrial Exhibition, except the judges said he was too young to qualify. In 1842, he moved to Paris, where, four years later, he patented his saxophone designs: seven pairs of them in different sizes and tonal ranges. Later, a tangle of lawsuits over his patents led to repeated bankruptcy, and he died penniless in Paris in 1894, an unworthy ending for the man who gave us the magic of the sax.

If this were a podcast, I'd end with Lenny Pickett playing the old *SNL* closing theme: Howard Shore's "Waltz in A." Tenor sax, of course.

50

ALICE IN (PHOTOGRAPHY) WONDERLAND

A math teacher's hobby

In an age of smartphones and selfies, it's hard to imagine the challenges involved in creating a photograph in the pre-digital, pre-film era, when it took hours, not a fraction of a second, to capture an image. Between 1827, the date of the first surviving photograph (rooftops from a window in Bourgogne in eastern France by Nicéphore Niépce), and 1851, when the relatively efficient "collodion process" was invented, early photographers had two choices of medium: daguerreotypes and calotypes.

The *daguerreotype*, named after its French inventor Louis-Jaques-Mandé Daguerre, dates from 1839. While it can capture fine detail, it's a one-off process, and the resulting image needs to be viewed from just the right angle. The *calotype*, introduced in 1841 by Englishman Henry Fox Talbot, was more convenient than the daguerreotype in that it allowed multiple positive images to be made from a single negative but lacked the fine detail of the "French" system.

Neither process satisfied the needs of a relatively unknown British sculptor, Frederick Scott Archer. Wanting to obtain new commissions by sending photos of his work to potential clients, his experiments went beyond one-off daguerreotypes and low-resolution calotypes, including experimenting with a recently invented substance, "collodion," a mixture of guncotton (nitrocellulose), ether, and alcohol. A viscous liquid, collodion has the property of drying rapidly into a clear, thin elastic film. As

Charles Dodgson, aka Lewis Carroll, took this collodion photograph of Alice Liddell, dressed as Tennyson's "Beggar Maid," in her parents' Oxford garden in early June 1857.
METROPOLITAN MUSEUM OF ART, GILMAN COLLECTION, PUBLIC DOMAIN

a skin-like dressing, it was first used extensively on battle wounds during the 1853–1856 Crimean War.

Archer's genius was to recognize that collodion could serve as a perfect matrix for light-sensitive silver salts, and in 1851, he perfected "wet collodion" photography. Like its predecessors, the process was still complicated, delicate, and time consuming, but its advantages were immediately obvious to photographers of the time: high resolution, multiple copies, and exposure times of less than a minute, which was short in those days.

One of the earliest adopters of the wet collodion process was an Oxford mathematics teacher and polymath, Reverend Charles Dodgson, better known to the world as Lewis Carroll of the "Alice" books. In the space of 24 years, from 1856 until 1880, Dodgson made around 3,000

Selfie of Lewis Carroll (Charles Dodgson), 1832–1898, taken on June 2, 1857. NATIONAL PORTRAIT GALLERY LONDON, PUBLIC DOMAIN

images, of which about 1,000 survive. He took photographs of landscapes, buildings, famous people (including Alfred, Lord Tennyson and European royalty), and—his métier—children.

One of these, Alice Liddell, was the daughter of the dean of Christ Church College in Oxford, where Dodgson lived and taught. Between 1857 and 1870, he made 11 portraits of Alice alone and a further seven of her with her siblings, all of which survive. The most famous of these, taken in June 1857, is of five-year-old Alice dressed as "The Beggar Maid" after Tennyson's poem of that name.

Five years later, on the "golden afternoon" of July 4, 1862, Dodgson and his friend Robinson Duckworth ("Dodo" and "Duck") would take Lorina, Alice, and Edith Liddell ("Prima," "Secunda," and "Tertia") on one of their frequent rowing trips. As usual on these "expeditions," Charles Dodgson made up fantastic tales to entertain the kids as he and Duckworth rowed upriver. On this singular occasion, however, Alice begged him to write down the stories for her.

And thus was born *Alice's Adventures in Wonderland.*

51

THE SOUND GUYS: EDISON AND BERLINER

Cylinders: 0; Discs: 1

Fifty years after the world's first surviving photograph was created by Nicéphore Niépce in 1827, Thomas Edison did for sound what the French experimenter had done for light: capture it. Edison made his first recording, "Mary Had a Little Lamb," in December 1877, and—like Niépce's original photo—Edison's recording was crude, imperfect, yet brimming with possibility.

The summer of 1877 saw Edison trying to build a machine that could transcribe the Morse code of telegraph messages into grooves on paraffin paper. By that fall, he realized that the same basic system might be able to record sound. He wrapped tinfoil around a cylinder that he rotated by hand while a stylus scribed a groove into the foil; the louder the sound, the deeper the groove. The stylus was attached to a diaphragm that responded to sounds focused by a conical horn. Somewhat to his amazement, he was able to record his own voice and play it back by reversing the process. "I was never so taken aback in my life. I was always afraid of things that worked the first time," he told a reporter.

Edison envisioned that his machine might be useful for dictation in a business setting, not realizing that the future of sound recording would be music. Today, of course, we take for granted the ability to listen to a Beethoven string quartet or Kendrick Lamar's latest hip-hop at the touch of a button, but before the invention of sound recording, you might have

Thomas Edison with an early phonograph, 1878. BRADY-HANDY PHOTOGRAPH COLLECTION, LIBRARY OF CONGRESS

heard a favorite symphony or concerto just three or four times in your lifetime. That's assuming you lived near a concert hall and could afford to attend a performance. Clearly, Edison's tinfoil (later wax) cylinders weren't going to take music to the masses because they couldn't be reproduced, they were short, and the quality was terrible. What was needed for commercial success was a way of capturing several minutes' worth of music with reasonable accuracy, then of making unlimited copies of the original without losing quality.

German American inventor Emile Berliner rose to the challenge. While Edison went on to improve the telephone and develop a viable electric lightbulb, Berliner devised the flat disc for sound reproduction, with grooves cut laterally (unlike Edison's vertically cut cylinder), a format virtually unchanged for the next 60 years. Berliner's system required separate machines to record (left in photo) and play back (right), unlike Edison's original cylinder instrument, which did both. Berliner's real breakthrough was figuring out how to create a sturdy master disc from which multiple copies could be made. Starting in 1886, he invented a method by which a stylus, following the wiggles of a moving diaphragm, etched a spiral line on a zinc disc coated with a thin layer of wax. After the wax was dissolved in acid, the inscribed line on bare metal became a playable groove. It was then a matter of "pressing," that is, converting the original into a durable master disc (the groove mirrored as a raised line), from which faithful copies—records—could be made.

While Berliner's basic design was to last until Columbia Records introduced the vinyl hi-fi LP (high-fidelity long-playing) record in 1948, the man himself turned his talents to other directions. After losing a series of lawsuits, including a messy patent battle with Edison, he found ways to improve cloth looms and acoustic tiles before embracing the young science of aircraft design, specifically helicopters. Berliner

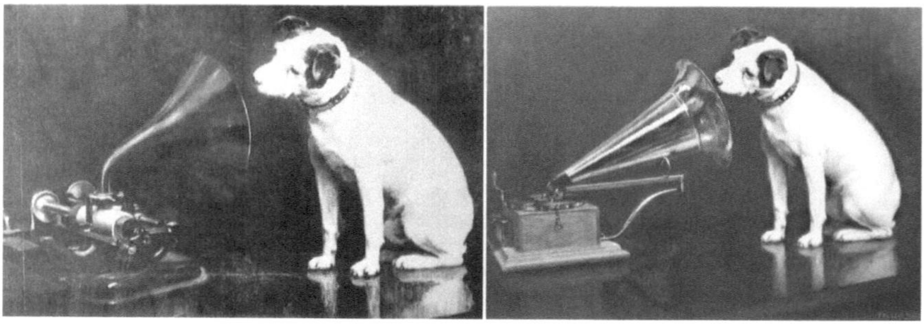

(Left) Francis Barraud's original painting with an Edison phonograph. (Right) The later one, in which Nipper is listening to His Master's Voice coming from a Berliner gramophone. WIKIMEDIA COMMONS

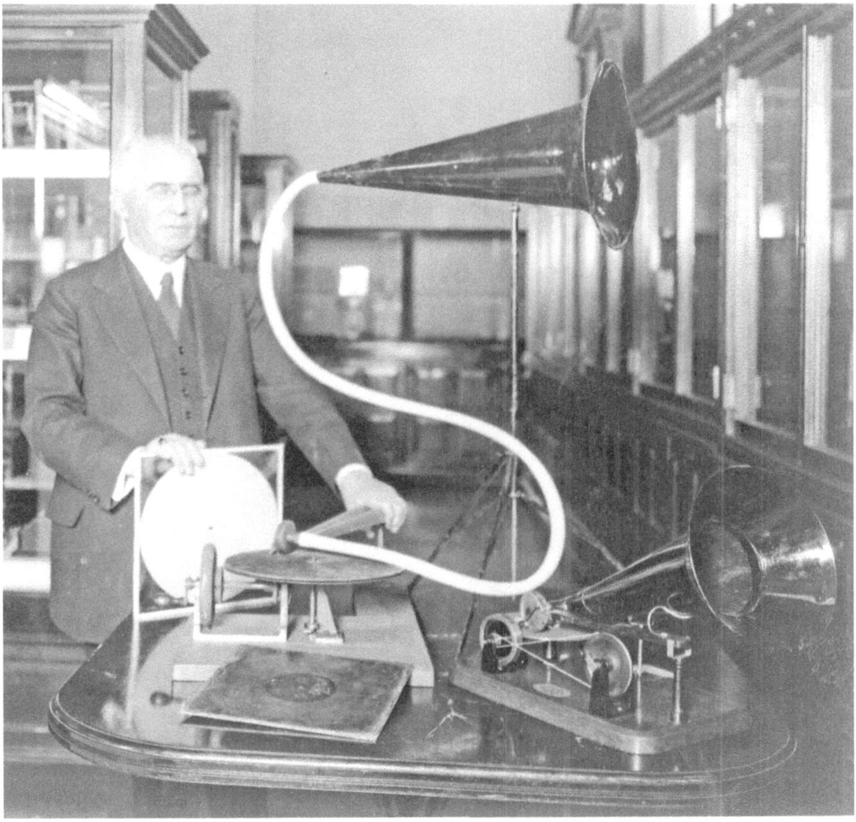

Emile Berliner (1851–1929) with his 1888 Gramophone on the right. Photograph taken around 1920. PUBLIC DOMAIN

introduced the use of the lightweight rotary engine (instead of a heavy in-line arrangement) and experimented with a vertically mounted tail rotor to counteract the torque generated by the main rotor.

So the next time you see an orange Coast Guard chopper, note the tail rotor and give a nod to Emile Berliner. But know that his real claim to fame is that he pretty much invented the recording industry.

52

THE GOLDEN AGE OF CRYSTAL SETS

That 1920s sound

Radio Sweeping Country: 1,000,000 Sets in Use
—*VARIETY* HEADLINE, MARCH 10, 1922

The history of commercial radio in the United States began when the Department of Commerce licensed station KDKA Pittsburgh on November 2, 1920, just in time for the Harding–Cox election results. Within a few years, hundreds of transmitters were broadcasting to millions of receivers, as virtually every household in America had either bought or built a simple crystal set. Those early sets needed no power source other than radio waves picked up by wire antennas (the longer and higher, the better), allowing farmers to hear the latest stock prices and Los Angeles sports fans to catch the first Rose Bowl on New Year's Day 1923.

I thought of this recently listening to a local talk-radio AM station here in Northern California. We live about two miles as the crow flies from the station's 5,000-watt transmitter, and, magically it seemed, that was enough to power my homebuilt crystal set via a 100-foot-long wire antenna strung outside our apartment. No batteries, no other source of power other than the radio waves themselves.

Here's how it works. An *antenna* picks up amplitude-modulated (AM) radio waves and converts them to electric current for the *tuned circuit*, which selects one signal out of all the radio frequencies received by

My homemade crystal set. Inset shows galena crystal with "cat's whisker." AUTHOR

the antenna by creating resonance between two energy-storing components: a variable capacitor tuner, which stores electrons within its dielectric insulator, and a coil inductor, which stores electrical energy as a magnetic field. When the resonance period matches that of the radio carrier wave, a signal is sent to the *rectifier* (or detector), which strips out half the radio wave, leaving a rectified radio-frequency (RF) signal. The original crystal sets were so named because their rectifiers were "cat's whiskers" (thin wires) touching crystals of galena (lead sulfide). Karl Braun discovered the rectifying property of certain crystals in 1874, allowing reception of "spark-gap" signals from transmitters hundreds of miles away. Today, hobbyist crystal sets use semiconducting germanium diodes, which are far more reliable than galena crystals. The rectified RF signal then goes to the *earpiece*. Moving-iron earphones were the most expensive component of a crystal set. Earphones converted the rectified signal into sound waves, filtering out the high-frequency "carrier" wave in the process. Hobbyists now use smaller and lighter piezoelectric earpieces. (Piezoelectric crystals bend when an electric field is applied, and, conversely, they generate a current when bent, as in gas-grill spark igniters.)

The demise of crystal sets began with the invention of "Audion" vacuum tubes (precursors of transistors), which amplified the weak signals sufficiently to power loudspeakers. The simple gadgets enjoyed a brief revival during World War II after Allied soldiers at Anzio were forbidden to use personal radio receivers because the Germans could pick up signals from the sets' superheterodyne oscillators. When inventive GIs realized that a pencil lead sitting on parallel razor blades made for a pretty good rectifier, "foxhole radios" were born.

The signal generated by a crystal set is tiny, taking full advantage of the incredible sensitivity (10^{-16} watts/cm^2) of the human ear. Devoted "xtal" aficionados on the West Coast can even pick up shortwave broadcasts from the BBC in London, a low-tech triumph in our high-tech era.

53

THE QUARTZ CRISIS

Adventures in watchmaking

The "quartz crisis" began in Tokyo on Christmas Day 1969, when Seiko unveiled the world's first quartz watch, the Astron 35SQ. Designed by Kazunari Sasaki of Suwa Seiko-sha, the Astron achieved 100 times the accuracy of a regular mechanical watch by using a vibrating quartz crystal instead of a balance wheel for timekeeping. Although the Astron originally cost ¥450,000, the same as a Toyota Corolla at the time, the price would drop over the next 30 years to where a perfectly serviceable quartz watch, keeping time to within five minutes per year, can now be had for $10.

While quartz technology brought accurate time within the reach of everyone, it would create a catastrophe for the Swiss watchmaking industry. Because Switzerland stayed neutral during World War II, the watch industry was unaffected, unlike other nations that shifted their watchmaking industry to timers for military ordnance. From 1945 until the early 1970s, the Swiss were able to capitalize on what was a virtual monopoly, accounting for more than 50 percent of the world watch market. All this changed with the advent of cheap and accurate quartz crystals; in the two decades following introduction of the Astron, the number of skilled workers employed by the Swiss watch industry fell from 90,000 to 28,000. What was a quartz crisis for the Swiss was a quartz revolution elsewhere, especially in Japan, where companies such as Seiko, Citizen, and Casio thrived.

Quartz crystals used in watches depend on the piezoelectric effect, discovered in 1880 by French physicists and brothers Jacques and Pierre Curie. Squeeze a crystal, and it will generate a tiny electric current; pass electricity through it, and it will deform. (The word comes from the Greek *piezein*, meaning "to squeeze.") Today, the piezoelectric industry accounts for $2 billion in annual business thanks to its use in inkjet printers, gas wands and lighters, guitar and drum pickups, microbalances, and atomic-scale microscopes.

If you're wearing a watch, chances are that inside, safe within a tiny capsule, there's a little quartz crystal (about one-tenth of an inch long) in the shape of a tuning fork. When a miniscule current is applied, it vibrates at a constant 32,768 times per second, or hertz. That frequency isn't just a random number—the size, shape, and plane on which the crystal is cut determines it. Why 32,768? It's 2^{15}. Halve it with a "flip-flop" circuit, and you get 16,384; do that 14 more times, and you get a once-a-second pulse.

The frequency of the quartz crystal largely determines its accuracy. The original Astron vibrated at 8,192 hertz (2^{13}) and was accurate to within one minute per year. Following the Astron, watches have incorporated higher frequencies (mostly 32,768 hertz) with a corresponding increase in accuracy to, perhaps, 10 seconds per year. Other key advances included replacement of the analog hands with a digital face, originally LED (light-emitting diode) and later LCD (liquid crystal display).

It took a while for the Swiss watchmaking industry to recover from the quartz crisis/revolution, but a glance at the advertising pages of any glossy magazine aimed at the upper crust will tell you what happened. The surviving firms (other than Swatch, another story) went for the top 1 percent of society with *very* fancy mechanical watches. We're talking thousands of dollars for a Rolex to $1 million and beyond for, say, a handmade Patek Philippe or Greubel Forsey (Google "tourbillon" for a glimpse into another world). I doubt their well-heeled owners wear them every day; I can imagine no greater invitation to being mugged.

Before the quartz revolution, even the best watches were hard pressed to maintain accuracy of 10 seconds per day, that is, one hour per year. With quartz, the manufacturers attained 10 seconds per year at best. *How about one second per year?* That's what's claimed for the Citizen Eco-Drive Caliber 0100, the world's most accurate autonomous watch. "Autonomous" here means it does this without reference to a radio or GPS time signal. One second per year is equivalent to an astounding one part in 31,556,952. One trick it uses is to have the quartz vibrate much faster than "regular" quartz watches, more than 8 million times per second.

That's quartz. There are, of course, many mechanical "chronometers" (adspeak for accurate watches; certified chronometers are something else) available. They come with serious price tags, hundreds of thousands of dollars for handmade ones. But when it comes to accuracy, mechanical watches, beautiful as they are, just aren't in the same league as quartz ones.

For example, the popular Rolex models (10,000 sold every week) are accurate to only a couple of seconds per day (10 minutes per year). In fact, most high-end manufacturers don't advertise the accuracy of their watches. You don't wear a $300,000 Patek Philippe to know what time it is. (You have an iPhone for that.) You wear it for its craftsmanship and its aesthetic appeal and, of course, to impress. One "watch blogger" was assured by a contact in Switzerland, "You aren't taken seriously in business in Europe if you don't wear a good mechanical watch."

54

ANALOG YOUTH TO DIGITAL GEEZER

Trying to keep up

The future isn't what it used to be.
—Yogi Berra, maybe

I'm about three years older than ENIAC, the first general-purpose electronic computer. ENIAC, which stands for Electronic Numerical Integrator And Computer, was a "Turing-complete" digital computer, that is, capable of being reprogrammed to solve virtually any numerical problem. ENIAC, which weighed nearly 30 tons, was the size of a small house, needed 150 kilowatts of electricity to run, and contained 17,468 vacuum tubes, 7,200 crystal diodes, 1,500 relays, 70,000 resistors, 10,000 capacitors, and millions of hand-soldered joints. It performed at a speed of 18 calculations per second and cost (in today's dollars) about $6 million to build in 1945.

Today, I'm editing this essay in a local coffee shop on a $349 Apple iPad. Weighs about one pound; smaller than a magazine; 10-hour battery life. Speed: more than 2 billion calculations per second.

I suppose every age looks back at a previous one and remarks on how far it's come. Phoenician scribes probably scoffed at old-fashioned cuneiform writing. The architects of the first gothic cathedral might have scorned the builders of those squat Norman piles of stone. I bet John Dunlop, bouncing along Belfast's cobbled streets on the pneumatic tires

ENIAC (Electronic Numerical Integrator And Computer) in Philadelphia sometime between 1947 and 1955. UNCREDITED U.S. ARMY PHOTOGRAPHER

he'd just invented, pitied bicycle riders of his dad's generation. But I do believe that the shift in my own lifetime from ENIAC to smartphone, from analog to digital, and from two-week international aerograms to WhatsApp and streaming videos represents more than a gradual, routine evolution of technology from one generation to the next. Hackneyed as it is, the phrase "quantum leap" really means something today. Change, for better or worse, is spiraling way beyond my ken.

Take *knowledge* in broad, raw terms. It used to be that one person could be the repository of a fair summary of the world's acquired wisdom. Leonardo da Vinci is sometimes cited as the original "renaissance man," although the same polymath sentiment might be applied to Aristotle, Francis Bacon, Thomas Jefferson, Johann von Goethe, Alexander von Humboldt, or H. G. Wells. No longer: too much knowledge, too many areas of specialty, too small brains. As a comparison, Google processes *10 times* the 2.5-petabyte (2.5 times 10^{15}) memory capacity of the human brain *every day*!

What's a sane reaction? Shock and awe? Humility? Fear? Another mojito? My own response to the threat of data overload—coupled with exponentially increasing processing power—is to surrender some of my need to know. Instead, I find myself simply enjoying the questions, of which there are plenty. According to the late "grand old man" of physics John Archibald Wheeler, just three problems will keep us happily occupied for the next 500 years: the *universe*, *consciousness*, and the *quantum*. I'd settle for simply (!) posing the right question, never mind the answer, for any one of these.

55

CHARON'S OBOL, THE ZAHIR, AND MORE

Riffing on coinage

My luggage is but a flask, a wallet, my old cloak and the obol that pays the passage of the departed.

—Leonidas, third century BC

Today, we take coins for granted, but there was a time when being able to trade a small disc of metal for a sheep, a bushel of corn, or a night's lodging must have appeared as some sort of alchemy. The first coins, dating to around 650 BC, were minted in Lydia, an Iron Age kingdom located in present-day western Asia Minor. They were made of electrum, a naturally occurring alloy of gold and silver, like the one-third stater in the illustration, stamped with the royal lion. According to the *Oxford Handbook of Greek and Roman Coinage*, three of these represented about a month's pay for a soldier. All coins used around the world today, including the dimes and quarters jangling in your pocket or purse, are descended from Lydian coinage like this via ancient Greece and Rome.

One of the most important coins to the ancient Greeks and Romans, at least according to their myths, was "Charon's obol." Charon, the Hades ferryman who carries the newly departed across the river Styx dividing the living world from the dead, demands an obol for the crossing. This is why many skeletons found in ancient graves in the Mediterranean world are found with a coin in their mouths—traditionally an obol, or one-sixth of a drachma, worth about $10 today.

Top: Lydian one-third stater, ca. 620 BC, worth around $3,000 as a collector's item today. Below: A 2019 U.S. one-ounce $50 Gold Eagle, worth about $5,500 in early 2026. GNU FREE DOCUMENT LICENSE

More recently in the literature, whale-obsessed Captain Ahab nailed a one-ounce gold "sixteen-dollar piece" to the mast of the *Pequod*. "Look ye! d'ye see this Spanish ounce of gold? . . . Whosoever of ye raises me a white-headed whale with a wrinkled brow and a crooked jaw . . . he shall have this gold ounce, my boys!" The coin, we learn later in Melville's *Moby Dick*, is an Ecuadorian eight-escudo doubloon, minted in Quito between 1838 and 1843.

In the United States, the U.S. Mint has produced gold coins since 1795, other than a 53-year break from 1933 to 1986 as a result of the Gold Recall Act. Currently, the mint produces the Gold Eagle in one-, one-half-, one-quarter-, and one-tenth-ounce sizes. They actually weigh a little more than the specified weights, as they're 22-karat (not 24, that is, 100 percent) gold, due to the addition of silver and copper for wear resistance.

And then (shiver) there's the Zahir. This is the 1929 Argentine 20-centavo coin given in change to the protagonist of Jorge Luis Borges' short story *The Zahir*, whose name happens to be Borges. It's a coin then, in 1949, but in previous incarnations, the Zahir has also been a tiger, a blind man, an astrolabe, a compass, a vein in one of the 1,200 marble pillars of the Cordoba Mosque, and the bottom of a well. If you're already a Borges fan, this litany won't surprise you. If not, you have a treat coming. The Zahir—one of the 99 Names of Allah ("The Manifest")—is "a person or an object that has the power to create an obsession in everyone who sees it, so that the affected person perceives less and less of reality and more and more of the Zahir" (Wikipedia).

Borges—the fictional one, we trust!—realizes that his fate is madness as he becomes completely consumed by the coin. "Others will dream that I am mad, and I [will dream] of the Zahir. When all men on earth think day and night of the Zahir, which one will be a dream and which a reality, the earth or the Zahir?" Read at your peril!

56

MASTODONS IN GREENLAND

How an invention helped re-create an ancient land

At first blush, the headline made it sound like something out of science fiction: "Scientists re-create ancient ecosystem by studying dirt!" But it's real. The dirt in question, 41 sediment cores taken by Danish paleontologists from the 2-million-year-old northern Greenland permafrost in 2005, turned out to contain a treasure chest of DNA. They didn't know that back then—the technology wasn't up to the task of teasing out that information—but on an inspired hunch, they stored the cores in a basement freezer at the University of Copenhagen while the extraction methods and sequencing devices improved incrementally. So every couple of years, they tried, unsuccessfully, to extract DNA from the Greenland samples using the latest equipment and know-how.

Finally, following a major upgrade in their instruments in 2017, they triumphed. After sifting and sampling some 16 billion "reads" (short fragments of DNA), they were able to identify traces of hundreds of animals and plants, many of which are now extinct. The trick was to match the reads—many as short as 50 base pairs—against entire known genomes, which typically run to a billion or so base pairs. Amazingly, that was sufficient to pinpoint which plant or animal the read came from.

Until this technological breakthrough, the oldest known DNA (present in every cell of every living organism) came from a 1.2-million-year-old mastodon molar. That was exceptional. Enzymes usually break down DNA within a few hundred thousand years, but in the case of the

Two million years ago, a warmer climate in northern Greenland, 600 miles south of the North Pole, supported a rich ecosystem of flora and fauna. The artist's impression is based on snippets of DNA extracted from permafrost in what today is a harsh polar desert. BETH ZAIKEN, BASED ON WORK FROM ESKE WILLERSLEV AND KURT H. KJÆR OF THE UNIVERSITY OF COPENHAGEN, USED WITH PERMISSION

Greenland cores, DNA—from sloughed-off skin, poop, or rotting leaves, for example—had bound itself, electrostatically, to minerals in the soil. (DNA has a weak electric charge, as do grains of clay and quartz in the sediment.) The telltale fragments of DNA were thus protected over the millennia from decomposition. Ironically, this same electrostatic attraction made the researchers' task more challenging, requiring innovative techniques to unbind the precious "reads" from the surrounding sediment.

Using the new technology, Danish scientists have been able to identify more than 100 kinds of plants, many of which had been found previously only as fossils, proving that today's polar desert was once a forested estuary dominated by birch and poplar trees together with sedges, horsetails, willows, and spruce. Several of the still-extant plants no longer grow in Greenland, indicating that the climate back then was warmer. And not only plants: the researchers have now detected DNA from several dozen species of living or extinct animals, including hares, reindeer, horseshoe

While most groups of the last mammoth species, the woolly mammoth, went extinct around 11,500 years ago, some isolated populations survived much longer. A herd living on Russia's Wrangel Island probably survived until about 4,000 years ago. © BETH ZAIKEN, USED WITH PERMISSION

crabs, rodents related to modern lemmings, precursors of today's geese, and—incredibly—mastodons, extinct relatives of elephants.

This new technique is revolutionizing the field of paleontology. We now know it's possible, given the right circumstances—in this case, suitable minerals for the DNA to bind to plus a cold climate—to develop a credible picture of an entire ancient ecosystem. The very limit for DNA survival is probably about 4 million years, a 10-fold jump from the few hundred thousand years to which researchers have previously been limited. In the future, university courses for budding paleontologists will no doubt include the arcana of microbiology, including DNA recovery and sequencing, as this new science transforms our knowledge of the past.

LANGUAGE

Words, spoken and written

57

LANGUAGE: NATURE OR NURTURE?

Or, is Chomsky irrelevant?

The linguist who does not display at least token interest in the Chomskyan endeavor is not considered a linguist, but instead perhaps a folklorist or an anthropologist who happens to study speech.
—John McWhorter, *The Language Hoax*

Many years ago, while living in Boston, I started to worry that I'd inadvertently joined a cult. I was taking an introductory linguistics class, and our teacher, a distinguished Harvard professor of language, intoned the name "Chomsky" so frequently and with so much reverence that I wondered if I might have stumbled into a church of that name. In 1978, the "Chomskyian revolution" was in full swing, and the proposal by Noam Chomsky, "father of modern linguistics," that we are born with an innate "Universal Grammar" (UG), had been accepted as gospel truth by many, perhaps most, professional linguists.

I bought it, hook, line, and sinker, and have since written approvingly of his theory over the years. Now, though, having followed the "language wars" for decades, I've recently become increasingly skeptical of UG. First, though, let's be clear on what Chomsky is claiming.

In a nutshell, language is innate to humans, and we are born "prewired" with a UG to spontaneously speak whichever language we hear in the first months of our lives. That is, our brains are equipped—at least metaphorically—with some sort of "language organ," the hardware on which the software of our parents' and tribe's language runs. When Chomsky introduced the notion of pre-wiring in the 1950s, he said that a Martian linguist would consider all 6,000 of Earth's languages to be variations on a single UG. Deep down, he says, all languages work the same because we come into the world with a similar capacity for language built in.

RECURSION

One of the consequences, according to Chomskyites, is that a fundamental feature of each of all 6,000 of Earth's languages is "recursion" (later renamed "merge" in a slightly diluted form), which governs how words are combined into more complicated phrases, with objects referring back to themselves within a sentence: "I saw the guy you said wore the same hat you bought from the shop we went to yesterday next to the market where . . ." (think matrioshka dolls). In a paper he cowrote for *Science* in 2002, Chomsky claimed that recursion is "the only uniquely human component of the faculty of language."

Finding just one language in which this feature is absent would negate Chomsky's "universal" claim. Linguist Daniel Everett (now at Bentley University in Waltham, Massachusetts), in his analysis of the Amazonian Pirahã language, claims to have done just that. Everett's 2008 book *Don't Sleep, There Are Snakes* is a terrific read both for his account of 20 years living with the Pirahã tribe and for his succinct repudiation of Chomskyian theory. Sadly, instead of considering Everett's field research at face value, Chomsky chose to retaliate with personal attacks, calling him a "charlatan." So much for academic professionalism!

According to Everett, rather than being innate, language is a human invention, like stone axes or huts. "The discovery that humans are better at building human houses than porpoises tells us nothing about whether the architecture of human houses is innate," he wrote, leading the charge against Chomsky's claim of universality. Everett, who was a loyal Chomskyite until 2000, is an unlikely antagonist. In 1977, he went as a missionary to live with the Pirahã, a tribe of 300 or so hunter-gatherers living in the Brazilian rainforest on the Maici River. His original intention

Left: Noam Chomsky in 2016. AUGUSTO STARITA, WIKIMEDIA COMMONS. Right: Daniel Everett and Kaabohoá, a Pirahã friend. MARTIN SCHOELLER, USED WITH PERMISSION

was to convert them to Christianity, but they were too happy to have any interest in being "saved." Years of observing their joy-filled live-for-the-day attitude resulted in *his* conversion to atheism! Meanwhile, he struggled to learn their esoteric language (three vowels, eight consonants, present tense only, no number or color words, 65,000 possible verb forms). Today, he is probably the only outsider completely fluent in Pirahã.

Everett found that the Pirahã language completely lacks recursion, mirroring the tribe's limited worldview. Anything outside their immediate experience—that they haven't either seen for themselves or heard from an eyewitness—is simply not worth discussing, meaning that their range of conversation topics is constrained, and hence there's no need for recursion. It follows that their sentences have finite length (whereas in Chomsky's view, sentences can be extended indefinitely). To use Everett's example, you can't say "John's brother's house" in Pirahã. Instead, you say (sentence 1) that John has a brother and (sentence 2) that this brother has a house. If true, this lack of recursion in Pirahã threatens not only Chomsky's claim that every language has recursion but also the entire edifice of UG.

Chomsky's original (1957) proposal is that we're all born with a kind of language module in our brains and that each of us, wherever we were born, adapted at a very early age to the language in which we were raised. This was his solution to the apparent incompatibility between (a) the limited speech data that young children throughout the world hear from their parents and caregivers ("poverty of stimulus") and (b) their near-perfect verbal ability by the age of six or seven.

IRREGULAR ENGLISH

In Chomsky's view, taking English as his example, this language is so irregular as to defy learning by trial and error. Consider, for example, the oddball rules in English that adult students of English as a second language have to struggle with:

- Irregular verbs, such as "said," "wrote," "swam" (not "sayed," "writed," "swimmed").
- Transitive and intransitive verbs: "We found the room" but not "We arrived the room."
- Declarative-to-question transformation: "The dog is there" becomes "Where is the dog?" That is, "there" transforms to "where" and moves to the front of the sentence.
- Copulas, such as "do," in the question "Do you mind?" (The vast majority of languages skip this "do" form.)
- Multiple auxiliaries.

Taking the last one, the "multiple auxiliaries" construction is one mainstay of UG. Consider the sentence "The dog in the corner is hungry" (using an example from Chomsky and elaborated by David Shariatmadari in his book *Don't Believe a Word*). To turn that into a question, we move the auxiliary "is" to the front of the sentence, "Is the dog in the corner hungry?" So far, so good. Now add a second auxiliary: "The dog that is in the corner is hungry." Move the first auxiliary to the front, and the question becomes the odd-sounding "Is the dog that in the corner is hungry?" Children rarely make a mistake like this. They figure out, early on, to say, "Is the dog [that is] in the corner hungry?" That is, they correctly move not the initial "is" but the one in the main clause to the front, dropping the secondary "is."

How do they learn to do this? According to Chomsky, "You can go over a vast amount of data of experience without ever finding such a case"—the case being "question" sentences like this with two auxiliaries. This is a core finding of his "poverty of stimulus" argument. Kids never hear such a question, he says, so they must have been born with the rule somehow embedded in their genes. Similarly, Steven Pinker, whose best-selling 1994 book *The Language Instinct* brought linguistics to a whole generation of curious minds, makes the same "poverty of stimulus" claim. He writes, "Questions with a second auxiliary embedded inside the subject phrase [like the 'dog in the corner' example above] are so rare as to be non-existent in Motherese."

It turns out that they're both wrong and that such sentences are two-a-penny! Researchers who recorded endless hours of Pinker's "Motherese" concluded that double auxiliary questions are frequently used with children. Kids don't need a "language module" in their brains to learn this; they just have to listen. "Where's the other doll that goes in here?" and similar questions are common in children's play environments.

NATURE VERSUS NURTURE

Where does all this leave Chomsky's UG? Without two of his key arguments—recursion and poverty of stimulus—there's not much left of his "nature" thesis. But let's not throw the baby out with the bathwater. Coming back to the question "Is language 'nature' or 'nurture,'" that is, "Is it innate or learned?" the answer is yes. That is, it's a false dichotomy (like virtually every other aspect of human behavior!). Of course, "nurture" is involved—that's why we speak the language of our parents or caregivers. But "nature" has to play a role too. Lacking *any* genetic learning module, we'd never pick up language in the way that everyone on Earth does, early and effortlessly. As Pinker (pretty much a Chomskyite) puts it, "There is no learning without some innate mechanism that makes learning happen."

In the case of language, the common machinery we're born with seems to result in at least some universals of language, such as nouns and verbs, cases and auxiliaries, and so on. The real question—what the "language wars" are all about—is this: do we have a *specific* module devoted uniquely to language (Chomsky's claim), or do we have a *general* skill-learning module that our ancestors co-opted for language?

According to Everett, it's the latter. Language learning can take its place in the treasure trove of universal—so presumably innate—human traits that include smiling as a friendly greeting, flirtation, the taboo on mother–son incest, reciprocity and retaliation, property inheritance, or having sex in private. (Anthropologist Donald Brown posited no fewer than 211 such universal characteristics.) What language, more than any other trait, demonstrates is the flexibility and potency of our capacity to learn. Not quite magic but pretty close!

This all may sound pretty arcane, but the argument has been the stuff of the high-octane language wars that have been waged, vociferously and ferociously, for decades. If anything, the debate is muddier than ever, with no sign of being completely resolved.

Or maybe (my own view, for what it's worth) it's just fading from relevance as a new generation of linguists mostly assumes that language acquisition is one feature of our general ability to learn skills—without accepting Chomsky's more extreme claim that we're born with a specific capacity for grammar. As Ted Gibson, professor of both cognitive sciences and linguistics at MIT, put it, "It's crazy to say [UG] is dead. It was never alive."

58

QUEST FOR OUR MOTHER TONGUE

Making sense of "Proto-World"

According to most linguists, all languages descend from a single mother tongue, dubbed "Proto-World," "Proto-Sapiens," or the "Ur" language. Invented maybe 100,000 years ago—estimates vary wildly—before humans left our African homeland, it spawned about a dozen "proto-languages" from which the world's 6,000 living languages (and countless dead ones) emerged. Proto-Indo-European (PIE), for example, was spoken around six or eight millennia ago on the steppes north of the Black Sea before diverging into a bevy of tongues, one of which you're reading now.

The strenuous and only half-successful efforts by linguists at reconstructing PIE and its progeny over the past 200-plus years give a sense of how difficult it is to re-create a long-dead language. PIE sired 10 well-attested "clades," or subfamilies, of languages (and several more disputed ones), including the three from which English is mostly derived: Proto-Germanic, Proto-Italic, and Proto-Greek. To reconstruct PIE, historical linguists look at common words in languages they suspect to be related, such as Spanish, Portuguese, Romanian, and Italian, and hypothesize rules for how they evolved from earlier roots. In the case of the Romance languages mentioned above, the task is, of course, made much easier by our extensive knowledge of their common ancestor, Latin. Usually, we don't have access to a written parent tongue.

The Confusion of Tongues, Gustave Doré, 1865. PUBLIC DOMAIN
Another version of how one language became many languages is found in the Book of Genesis. "Seeing the unchecked power of humans speaking one tongue, God commanded: 'Go to, let us go down, and there confound their language, that they may not understand one another's speech.'"

If this sounds complicated, it is. Linguists have taken more than 200 years (since 1786 with the proposal that Sanskrit, Greek, Latin, Persian, Gothic, and Celtic derive from a common root) to reconstruct PIE by comparing the lexicons, morphologies, and grammars of its many offspring. Even so, our knowledge is still very limited, and sounding out just a single well-formed sentence in PIE is still the subject of conjecture.

Given the difficulty of reconstructing PIE, a language spoken just a few thousand years ago, imagine how much more problematic it is to identify words from 100,000 or more years ago in Proto-World. For that daunting task, a few brave (or foolhardy, take your pick) linguists have taken a different approach to the "comparative reconstruction" method used to re-create PIE and other protolanguages. These mavericks use a controversial technique known as "mass comparison," that is, comparing hundreds of languages from around the world looking for commonalities without worrying too much about how they actually evolved.

RECONSTRUCTING PROTO-WORLD?

Starting in 1994, two of the most outspoken Proto-World linguists, Merritt Ruhlen and John Bengtson, published a list of 20-odd "global etymologies" for such common words as "hair," "water," "two," "finger/one," "who," and "what." According to their reconstruction, speakers of the mother language may have used *aqwa* for "water," *tik* for "finger/one," *pal* for "two," and so on.

Some of these—perhaps all—are a stretch. Take *aqwa*. Sure, we can see Latin *aqua* and a couple of other less obvious derivations, but Ruhlen and Bengtson claim, for example, that the words for "water" in various protolanguages, including *nki*, *engi*, *niru*, *namaw*, and *gugu*, are related. For the vast majority of comparative linguists, extrapolations like these are just too much to accept. For that matter, it beggars the imagination to suppose that *aqwa* could have survived unchanged for 100,000 years!

One outspoken linguist who takes exception to Ruhlen and Bengtson's proposed etymologies is American linguist John McWhorter, who has written extensively on Proto-World. As his go-to example of rapid language change, McWhorter discusses the Cheyenne word *aa* for "winter." Around 1,500 years ago, speakers of Cheyenne's precursor language, Proto-Algonquian, were using *peponwi* for "cold season." Historical linguists have been able to show exactly how a well-documented series of sound changes (*peponwi-pepon-eon-ain-ai-aii-aai-aa*) morphed *peponwi* into *aa* in just 1,500 years. Yet the Proto-World reconstructionists claim to have derived words from nearly 100 times as far back, hence the rather cruel dismissal of their efforts by one mainstream researcher as the "cold fusion of linguistics."

PIDGINS AND CREOLES

So reconstructing any actual words used by the first speakers is probably a lost cause. However, we can still make some intelligent guesses at what *kind* of language, in very broad terms, they used. Consider the closest analogies we have to "virgin languages" today: *creoles*. A creole starts out as a pidgin language, that is, a highly simplified method of communication between two groups of people who lack a common language. For example, when whites encountered Australian aborigines in the late 1700s, both parties figured out a pidgin English for trading. For whites and aborigines alike, this was a second language, learned as adults.

However, when children hear adults speaking a pidgin and start speaking it among themselves as their primary tongue (in a process known as "nativization"), it morphs into a creole, with at least some of the subtleties inherent in any natural language. So while the *vocabulary* of a creole derives mostly from the pidgin's parent language(s), its *grammar* and *syntax* arise spontaneously and naturally. But here's the odd thing: *whatever the structure of their parent languages, creoles the world over have much grammar and syntax in common.*

Virtually all creoles use subject-verb-object word order ("dog bit man"), are nontonal (unlike most languages today), ignore definite and indefinite articles ("the" and "a"), do not differentiate between verbs and nouns (context supplying the meaning), and lack tenses ("She go yesterday.").

Isn't this astonishing that when humans—in this case, children of adult pidgin speakers—unconsciously decide on the syntax and structure of language, they choose the same universal forms?

According to many linguists (but not all—they tend to be a combative bunch), our Proto-World mother tongue likely had similar forms to that of first-generation creoles. We'll probably never know if our distant ancestors used *tik* for "finger" or *aqwa* for "water" or any other specific words. But by listening to speakers of creole today, we can get a sense of what those inventive first-language speakers might have sounded like.

59

THE WEIRDNESS OF ENGLISH

Black sheep of Germanic languages

English is a mutt of a language, "the bastard tongue of a bastard tongue," in linguist John McWhorter's phrase. He was addressing the fact that the language you're now reading isn't just the odd man out compared with its northern European siblings but also that their parent language, Proto-Germanic, is itself an oddball in the Indo-European family.

No one paid much attention to language families until a polyglot judge based in India, Sir William Jones, gave a now legendary address to the Asiatic Society of Calcutta in 1786. He noted that the similarities between Sanskrit, Greek, and Latin couldn't be coincidental and that they must derive from a common root. (More tentatively, he suggested that the Persian, Celtic, and Gothic languages were also derived from the same root.) His observation—not totally original but the first to tabulate and formalize the connection—kick-started the science of linguistics.

In a nutshell, around 6,000 years ago, nomadic Yamnaya people living on the steppes north of the Black Sea and the Caspian Sea (in present-day Ukraine) spoke a now extinct language we call Proto-Indo-European (PIE). They domesticated horses, used wheeled chariots, grew crops, and spun wool. Starting around 2500 BC, they spread out, east to India and west to Europe. Today, 2.5 billion people—one-third of the world—communicate using PIE-descended languages such as Urdu, Hindi, and Punjabi (in the Sanskrit family); Kurdish, Pashto, and

Sir William Jones (1746–1794), chief justice of Calcutta, spoke 13 languages fluently and knew 28 more reasonably well. His proposal that Sanskrit, Persian, and most European languages are cognate is now universally accepted. PORTRAIT BY SIR JOSHUA REYNOLDS, PUBLIC DOMAIN

Farsi (Iranian); French, Spanish, and Italian (Italic); Welsh, Breton, and Gaelic (Celtic); Polish, Czech, and Russian (Slavic); German, Swedish, and English (Germanic); and some 30 more distinct tongues.

FROM PIE TO PROTO-GERMANIC

Proto-Germanic, the parent of English and about a dozen other languages now spoken mostly in northern Europe, has several features shared with neither its parent (PIE) nor any of its siblings. In addition to

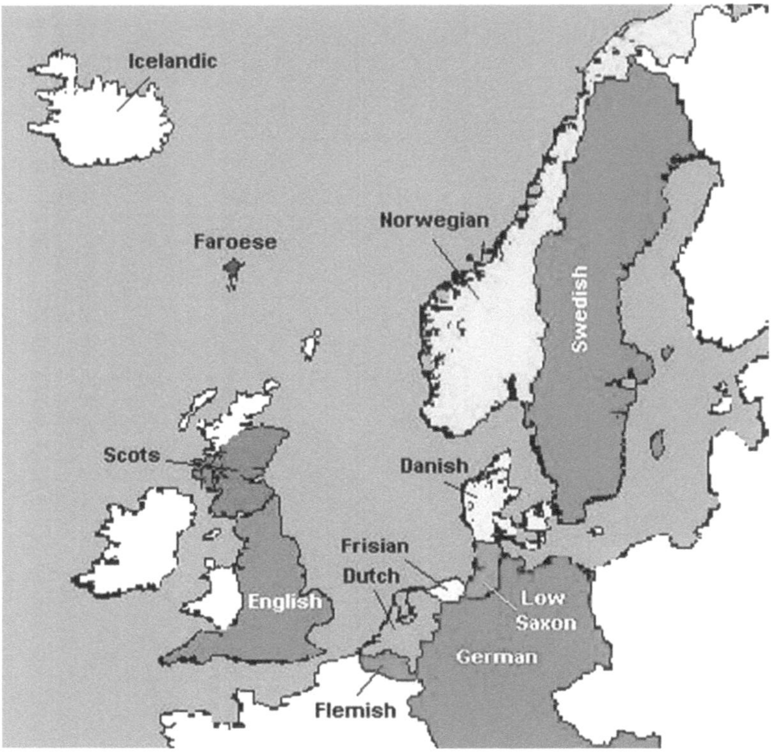

The most important languages derived from Proto-Germanic. KENNETH DOIG/ *NORTH COAST JOURNAL*, USED WITH PERMISSION

being much simpler, especially in its abbreviated verb endings and lack of genders, Proto-Germanic uses *fricatives*.

This is easier to show than explain by comparing Latin (a "regular" PIE language) to English:

- Latin *p* becomes English *f* (*pater* to *father*).
- Latin *t* becomes English *th* (*tres* to *three*).
- Latin *k* becomes English *h* (*canis* to *hound*).

The *p*, *t*, and *k* of Latin and all other PIE languages are *voiceless stops*. Try saying "p," "t," and "k," noting that your lips, tongue, and glottis momentarily stop the airflow. Now do the same with the fricatives (as in "friction") "f," "th," and "h." They're also created by your lips, tongue, and glottis, but in this case, you only partially stop the airflow.

What caused this "voiceless stop to fricative" switch in Proto-Germanic? The generally accepted hypothesis is that it was due to speakers of Semitic languages (which are awash with those sibilant *f*, *th*, and *h* sounds) and who learned PIE before passing it on, imperfectly, to their children. This can also explain the simplified verbs and nouns of Proto-Germanic, typical of adult second-language speakers. There's good reason to believe that these parents were Phoenician seafarers (see "The Phoenician Connection"). Meanwhile, speakers of other PIE-derived tongues passed on their language in the usual way, that is, native-speaking parent to child, thus retaining PIE's original voiceless stops and complex verb and noun forms.

Proto-Germanic, then, is uniquely odd. But so is one of its descendants, English, the black sheep of the Germanic language family. For one thing, its grammar is far simpler than any of its siblings, being the only language spoken in Europe that lacks gender (other than one dialect of Swedish). Whereas German, for example, has *der*, *die*, and *das* to signify masculine, feminine, and neuter, respectively, English has plain old *the*. English probably lost its gendered nouns in the transition between Old and Middle English as Danish and Norwegian newcomers struggled to learn "Englisc," streamlining it in the process.

IS ENGLISH A CREOLE?

Some linguists have even questioned whether English should properly be classified as Germanic in the first place given that its vocabulary is composed mostly of Romance languages (29 percent French and 29 percent Latin) compared to 26 percent Germanic (Old and Middle English, Old Norse and Dutch). One way of thinking about English is as a creole, created in the guise of Old English by Celtic-speaking Britons who learned it as a second language from successive waves of European invaders and settlers between about AD 450 and 1100. (Celtic is another branch of the Indo-European language family, represented today by Welsh, Breton, and Irish and Scottish Gaelic.)

Although Old English has traditionally been linked with Angles and Saxons (hence Anglo-Saxon) and Jutes, more recent studies indicate that it was essentially a mixture of Frisian (the closest relative of English, spoken in the Netherlands, where it's an official language) and Saxon (a hybrid created by British Celtic-speaking natives). And it's in that Celtic connection where the weirdness of English really shows. See "No Celtic? Blame the Volcanoes."

60

NO CELTIC? BLAME THE VOLCANOES

Why English lacks Celtic

Celtic is virtually absent in the English language despite the fact that almost everyone in what is now England spoke one or another version of Celtic before AD 410, when the departure of the Roman legions was a signal for Germanic speakers on the European continent to begin their centuries-long process of invasion. I say *virtually* absent: the exception is our use of the auxiliary "do" ("Do you see that?" "No, I don't.") and our "-ing" ending ("I'm writing," not "I write"). Of all the world's 6,000 languages, the only other two that use equivalent "do" and "-ing" constructions are Celtic Welsh, plus a near-extinct branch of Italian.

The near-total disappearance of the Celtic language in England is odd if you look at what happened in other Roman provinces, notably Gaul (France), Iberia (Spain and Portugal), Dacia (Romania), and Italy itself. There, the Germanic invaders—and ultimately settlers—adopted the language of the thoroughly Romanized natives. For example, by the time they conquered England in 1066, Normans ("north-men") had traded their Scandinavian tongue (Germanic) for French (Latin).

The case of England in the fifth and sixth centuries is more akin to what happened 1,000 years later, when European imperialists encountered Native Indians in the Americas. Instead of intermarrying with the locals, Germanic-speaking invaders from northern Europe—Angles and Frisians along with Saxons and Jutes—barely assimilated with native

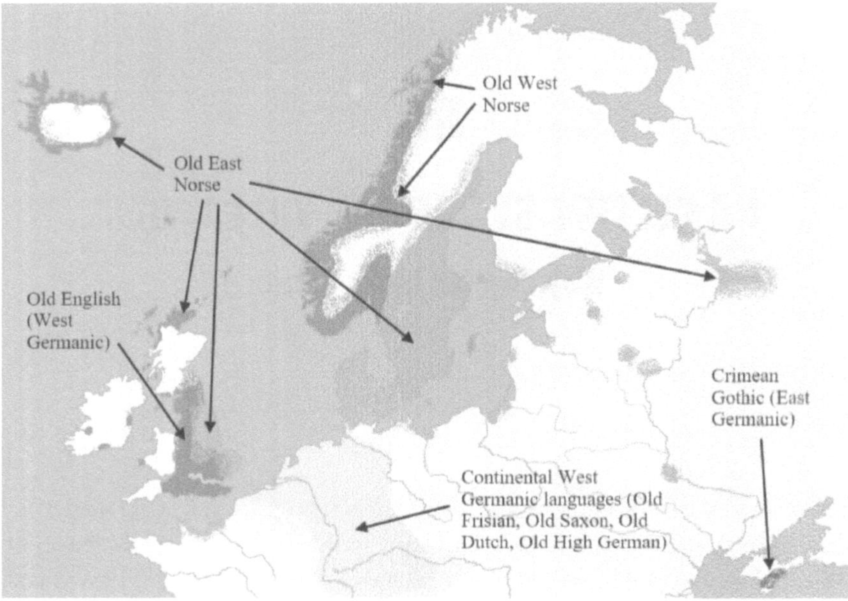

Germanic languages around AD 900. Celtic was spoken in the remaining (white) parts of the British Isles. ADAPTED BY THE AUTHOR FROM WIKIPEDIA

Celtic-speaking Britons. Instead, they drove the native population westward and northward. Archaeological (graves) and linguistic (place-names) evidence shows that by the early sixth century, the country was essentially split in two: Celtic speakers in the west, Anglo-Frisian (aka "Old English") speakers in the east.

PLAGUE!

Speaking mutually incomprehensible languages, the two groups had little in common. While the native Britons maintained connections to what remained of the Roman Empire via ports in the southwest of the country, the newcomers traded with their northern European homelands. So when the devastating bubonic plague ("Plague of Justinian") of AD 541 reached Britain around 547 (probably on ships from Rome via ports on the English south coast), it infected and killed Celtic speakers in the west of England to a much greater extent than it affected Germanic speakers in the east. Con-

Present-day outposts of Celtic languages. AUTHOR

sequently, the Angles, Saxons, and the rest were soon able to migrate west unopposed and settle the depopulated "wastelands" of Arthurian legend.

The plague came on the heels of devastating volcanic eruptions. The volcanoes haven't been definitively identified, but geologists believe that a terrible winter of AD 536–537 saw at least three simultaneous eruptions, leading to what some historians have labeled "the worst year ever." The resulting sulfuric aerosol "veil" led to at least a decade of global cooling during which "the sun gave forth its light without brightness, like the moon," according to one contemporary account. The resulting cold triggered crop failures and livestock deaths on a massive scale worldwide.

One of several climate-related factors that trigger plagues is temperature drop. The weather has to fall below 20°C (68°F) for an epidemic to break out because that's the temperature at which the flea that feeds on humans, *Pulex irritans*, becomes active. During feeding, plague bacilli (*Yersinia pestis*) in a flea's gut are passed into the bloodstream of its human victims.

So that's the theory: massive volcanic eruptions in late 536 led to the outbreak of bubonic plague, which, in 547, spread to Celtic Britons living in the west of Britain while mostly sparing the Germanic speakers in the east. This left all of England to the northern European invaders and settlers, which is why we speak a West Germanic language today that is nearly—but not quite—untouched by Celtic. All that we're left with is "do" and "-ing."

Which is why you can ask of the above, "Does he know what he's talking about?" but not "Knows he what he talks about?"

61

THE PHOENICIAN CONNECTION

How English "softened"

Grimm's Law was a breakthrough in the young science of linguistics when Jacob Grimm and his fellow Dane, Rasmus Rask, first formulated it in 1822. (We know Jacob and his brother Wilhelm for their collections of European folk stories that—in their cleaned-up versions—have kept Disney in business for decades.) The law, the first to codify systematic sound changes, noted the connections between Germanic language consonants (such as English) and their Latin equivalents. Proto-Germanic and Latin are two of several daughter languages of Proto-Indo-European (PIE), which was spoken by nomadic Yamnaya people around 5,000 years ago in the Balkans/Black Sea region.

According to Grimm's Law, the original PIE "voiceless stops" regularly and systematically changed to Proto-Germanic "fricatives" (see "The Weirdness of English"). Again, Latin *p* becomes English *f* (Latin *pater* becomes English *father*), Latin *t* becomes English *th* (Latin *tres* becomes English *three*), and Latin *k* becomes English *h* (Latin *centum* becomes English *hundred*). Note that *p*, *t*, and *k* sound "hard" compared to their *f*, *th*, and *h* equivalents because those voiceless stops involve an actual break in airflow not shared by *f*, *th*, and *h*.

Linguists have only recently suggested a convincing reason for Grimm's Law and why it isn't universal, applying to all languages derived from PIE. The odd thing is that Proto-Germanic is the only daughter language of the original PIE parent to exhibit this "voiceless-stop-to-fricative" switch. In fact, we can generalize the law to say that if in other Indo-European languages there is, for example, *p*, you'll find an equivalent *f* in Germanic. So while we have (non-Germanic Indo-European) Ancient Greek *pous*, Latin *pes*, Sanskrit *pada*, Russian *pod*, and Latvian *peda*, we see (in Germanic languages) English *foot*, Frisian *foet*, German *Fuss*, Icelandic *fotur*, and Danish *fod*.

Why? What happened to the Germanic languages that didn't happen to the rest of the Indo-European ones? Maybe we should blame (or thank) the Phoenicians. The idea is that when speakers of Semitic languages like Phoenician (which are rife with the sibilant *f*, *th*, and *h* sounds) learned PIE from native speakers, they spoke it like all adult foreign-language learners: with an accent, in this case with their own "hissy" accent, which their children would have then learned from them, passing it on to their own children, etcetera. So unlike the rest of the PIE daughter languages, which were transmitted via "native-speaking parent to child," Proto-Germanic appears to have undergone at least one "parent-speaking-a-foreign-language-to child" transmittal.

Phoenicians are the mostly likely "second-language-transmittal" candidates because they were the best navigators, traders, and explorers between 3,000 and 2,000 years ago, when Proto-Germanic originated. We know Phoenicians gave us our alphabet. Now it seems fitting, if the theory holds up, that they're also responsible for most of our consonants.

62

LORD, IS IT ME? (OR, IS IT I?)

The quest for "proper English"

When I was a kid growing up in the United Kingdom, world maps in our schoolrooms still showed huge swaths of pink, glorifying the extent of the British Empire, on which, we were assured, the sun never set. We were fed the myth that Britain ruled the waves (along with much of the land until our comeuppance at Suez in 1956), and moviemakers of that era played along. Typical scene: white meets black somewhere in the Dark Continent. "I'm Doctor Farnsworth," the pith-helmeted explorer would say in his plummy Eton voice. "Me Moogoo," replied the awed native.

We'd smile, of course. Back then, using object "me" for subject "I" was seen as a sign of inferiority. The idea was that if someone knocked and you asked, "Who is it?" their reply, "It is I" or "It's me," would immediately establish the class of whomever (!) was outside. Pronouns came in two basic flavors, subject and object, and good heavens man, if you're going to get anywhere in this world, you'd jolly well better know which one to use.

English pronouns really are a bit of a dog's breakfast. We differentiate between subject and object with some: "I"/"me," "he"/"him," "she"/"her," "we"/"us," "they"/"them." But we don't with others: "It" takes care of them both, while "you" covers not only subject and object but singular and plural as well. Which is why "youse," "you-all/y'all," and "you-uns" have been called into service, hearkening back to the archaic forms, singular “thee/thou” and plural “ye/you.”

In theory, I/me is pretty clear, but in practice, it just ain’t so. Okay, maybe the late Queen Elizabeth said, “It is I,” and she was probably the last living native English speaker to start a sentence, “My husband and I . . .” The rest of us peasants have mostly moved on. Here’s the test: would you say, “Pat and I met at the coffee shop” or “Pat and me met at the coffee shop”? If you’re partial to “I,” check this out: “I and Pat met at the coffee shop.” That’s “proper” English if you’re committed to the I-subject/me-object rule. But trust me, even the Queen didn’t say, “I and Philip are going to Ascot.”

Here’s what happened. Three hundred or so years ago, self-styled grammarians promulgated certain rules of English—check Wikipedia’s entry for “Robert Lowth,” for example—basing their notions of what constituted “proper” English on Latin, whose grammar is mathematically logical. (And complicated. Seared into my brain are the Latin cases: nominative, genitive, dative, accusative, ablative, locative, and vocative.)

But English isn’t Latin. Unlike Latin, it’s *alive*, meaning it’s fluid and flexible, changing from one generation to the next. And in spoken English today, pronouns don’t always follow mathematical logic. Time was, “Amn’t I?” sounded okay, but today, the illogical (singular/plural disagreement) “Aren’t I?” is standard. Pronouns are, in the words of popular linguist John McWhorter, “playful little puppies” that jump the rails of formal grammar. If I were (another usage in transition) going for a job interview, I’d probably say, “My friends and I . . . ,” but once I got the job, I’d get off my high horse and use “me.” I bet you-all would too.

63

HOMER'S WINE-DARK SEA

Did the ancients see our colors?

The organs of colour and its impressions were but partially developed among the Greeks of the heroic age.

—W. E. Gladstone, *Studies on Homer and the Homeric Age*, 1858

In Homer's two epic poems, the *Iliad* and the *Odyssey*, the phrase "wine-dark sea" occurs dozens of times. Actually, it's more like "wine-looking" (*oinops*, from *oinos*, "wine," and *op-*, the root for "see," as in "vision"). The phrase isn't of interest only to scholars of ancient literature but has attracted an abundance of scientific interest. Was Homer simply playing with a nice metaphor? Was he color-blind? (Or totally blind? as one tradition has it.) Did a repeated "red tide" (red-colored marine algae) suggest the phrase to him? Or maybe it was the *wine* that looked vaguely aqua after it had been diluted with alkaline groundwater, which, Photoshop-style, decreased red and increased blue.

William Ewart Gladstone, 10 years before becoming one of Britain's most liberal and effective prime ministers, accepted the challenge. He was enamored with the *Iliad* and the *Odyssey*, calling them "the most extraordinary phenomenon in the whole history of purely human culture." In his 1858 opus *Studies on Homer and the Homeric Age*, he proposed that ancient Greeks lacked our rich sense of color and that they saw their world much as we see at dusk when vivid noonday hues have been diluted to tinted grays.

As one line of evidence, Gladstone cites Homer's curious color vocabulary: in the two poems, the words for "black" and "white" occur 170 and 100 times, respectively, while Homer's most common color word, "red," appears a mere 13 times. "Blue" appears not at all. In fact, there was no word for blue in the Greek of that time! (The word that later meant blue, *kuaneos*, was simply "dark" for Homer, as in "dark-browed Zeus.")

How to explain the ancient Greeks' apparent color deficiency? In an outside-the-box speculation, Gladstone proposed that humans had acquired their sensitivity to the color spectrum since the time of Homer only as a result of Lamarckian inheritance.

LAMARCKISM

This is the theory that an organism can pass on characteristics acquired during its lifetime to its offspring. It was current at the time of Gladstone's speculations (and would be until about 1880, when it was overtaken by Darwinian evolution). Lamarckism isn't true: if it were, to take an obvious example, Semitic males would be born absent foreskins after untold generations of their circumcised ancestors. But for Gladstone, Lamarckism was a mechanism by which essentially colorblind humans could evolve color sensitivity in the space of a few generations. He cited the invention of artificial paints and dyes, starting around 3,000 years ago, as the catalyst for kick-starting our forebears' sense of color.

Several nineteenth-century anthropologists adopted Gladstone's theory, noting that the living languages of "primitive" societies routinely lack the full spectrum of color words employed by "civilized" Europeans. Perhaps, they argued, our darker-skinned brethren share the ancient Greeks' putative colorblindness—conveniently forgetting (again taking an obvious example) that their offhand use of "black" and "white" for skin coloring misrepresented reality. (When was the last time you saw a really white or black person?)

According to Gladstone, it wasn't just ancient Greeks. Many languages confuse, for example, blue and green. To cite a few of literally dozens of examples, Sioux's single word for "grue" (that is, blue *and* green) is *toto*. Similarly, "blue" (sky) and "green" (grass) take the single word *glas* in Welsh, *sheen* in Kurdish and Pashto, *urdin* in old Basque, *quin* in traditional Chinese, and *ao* in Japanese. Do these "grues" reflect a color deficiency in these people? Well, no, language use isn't the same as sensitivity. Populations worldwide distinguish as many shades of green and blue as do Europeans, even if they don't have as many color words as their genteel counterparts back in Paris and Berlin.

In his book *Through the Language Glass*, linguist Guy Deutscher proposes a counterexample to this apparent blue-green "confusion," inviting us to imagine a nineteenth-century Russian anthropologist speaking to the Royal Academy of Sciences in St. Petersburg. He tells the astonished members that, on a recent visit to the British Isles, he discovered that the poor natives there confuse the colors *siniy* and *goluboy*, calling them both "blue," even though they can perfectly well distinguish between the two. In the visible light spectrum, Russian *sinty*, what we might call navy blue, clocks in at a wavelength of about 430 nanometers (billionths of a meter), while the wavelength of *goluboy* (sky blue) is about 480, for a difference of 50 nanometers. Green, on the other hand, centers around 530 nanometers. That is, the actual (measurable) range of wavelengths in the colors we collectively call blue is about the same as the difference between sky blue and green!

So just as English speakers lump Russian *siniy* and *goluboy* into blue, other languages lump our blue and green into their version of "grue." Here, surely, is the answer to Homer's "wine-dark sea." The ancient Greeks were perfectly able to distinguish between colors, just as all modern people can (at least those who aren't lacking one or more of the three types of color-sensitive "cones" in their retinas). The fact that a language lacks the words for subtle and not-so-subtle shades of color tells us nothing about the sensitivity of the speakers' eyes.

Homer knew darn well what the color of the Aegean was. In fact, after a few generous cups of red wine on Ithaka's rocky shore, I bet any of us might be guilty of the same mistake. We'd call it "poetic license."

64

INVENTION OF OUR ALPHABET

From ox head to letter A

Take a moment to appreciate what you're now doing, making sense of little squiggles on the page in front of you. It's easy to take for granted our wondrous facility to read and write, but of course, it wasn't always so. Before the invention of writing, people had no way of passing on knowledge or history other than memory, transmitted orally from one generation to the next.

When the ancient Greeks lost the art of writing during their "Dark Ages," from around 1200 to 800 BC, the legend of the Trojan War was kept alive by storytellers whose only records had been in the brains of narrators who preceded them. The rediscovery of writing meant that Homer could reproduce the stories in a form that no longer depended on an uninterrupted line of performers. Nearly 3,000 years later, we are heirs to that tradition.

One of the great landmarks in the history of writing was the emergence of the alphabet, whereby a single sound is represented by a single character. This was a vast improvement over previous syllabic systems, in which symbols represented *syllables* rather than individual *sounds*. The problem was the sheer number involved. For example, Mesopotamian scribes used around 600 cuneiform (wedge-shaped) symbols before the invention of the alphabet. It was only when writing was reduced to a manageable number of characters that literacy moved from an esoteric skill (known only to scribes) to the common ability we enjoy today.

Most languages have no more than 30 or 40 individual sounds. The insight that writing could be reduced to a small number of distinctive signs, one sign per sound, is credited to Akkadians and/or Sumerians living in Mesopotamia around 1700 BC. Scribes there wrote with sticks to impress cuneiform characters into lumps of soft clay that were then baked in the sun or in a fire. Tens of thousands of these cuneiform clay tablets have survived to the present day. One of these, found in 1948 in the ancient city of Uragit in Syria, is particularly useful to historians. It's

Cuneiform writing on a clay tablet dated to 1635 BC, now in the British Museum. It recounts the Babylonian flood epic of Atrahasis, one of the inspirations for the Noah story. AUTHOR

an "abecedary," that is, the letters of the (Ugaritic) alphabet, written out in order. When you learned your ABCs, you were following an ancient tradition for teaching children how to write.

ALPHABET

The next great revolution in writing took place not on wet clay tablets but on the smooth, dry surfaces of papyrus or dressed stone, as wedge-shaped characters gave way to a linear alphabet. The ancient Phoenician port of Byblos (whence the word *Bible*) was probably the laboratory where scribes created a linear 22-symbol, consonant-only alphabet sometime before 1000 BC.

Every account of how the alphabet came into being inevitably references the sarcophagus of Ahiram, king of Byblos (modern J'beil, Lebanon), which has been dated to around 1000 BC. This stone sarcophagus has been housed in the National Museum of Beirut since shortly after its discovery by French archaeologists in 1923.

(We're lucky to still have it. From 1982 to 1997, this historic treasure was sheathed in reinforced concrete while fighting raged all around, literally. The museum happened to lie on the demarcation line between opposing forces during Lebanon's 15-year civil war and was used as a troop barracks. Today, the sarcophagus takes pride of place just inside the entrance of the newly rebuilt and refurbished museum.)

The importance of the sarcophagus lies not with the king who was once interred inside it or in the ornate relief scenes on its sides but in the writing carved around the cover. The 38 words recorded here represent the earliest recorded inscription of any length in the Phoenician alphabet. They are a warning to anyone who uncovers the sarcophagus in the future: *may the throne of his kingdom be overturned, and peace and quiet flee from Byblos.*

Our alphabet has its origins in this Phoenician alphabet. Phoenician wasn't quite the first "linear" script (Ugaritic—from ancient Ugarit in present-day Syria—preceded it), but it's the script from which most modern alphabets are derived since it led to Aramaic (the lingua franca of the ancient Mediterranean world), which in turn was adopted by the Greeks, who then passed it on to the Etruscans. And when the Romans conquered the Etruscans around 400 BC, they adopted and improved the script of their subjects. From there, it's a (fairly) small step to our 26-letter alphabet.

Top: National Museum of Beirut. King Ahiram's sarcophagus is on left in the middle, circa 1000 BC. Bottom: Earliest known Phoenician writing on top of stone cover comprising 38 words. BOTH AUTHOR

Like modern Hebrew, the Phoenician alphabet was all consonants, so vowels had to be interpolated by the reader. This rarely led to confusion (mst nglsh wrds cn b ndrstd wtht vwls). The earliest vowels were added by early Aramaic speakers, using letter shapes already in use. For example, they took the Phoenician bull's head symbol for a glottal stop (the middle sound in Cockney *bo'el*, "bottle") and used it to indicate the vowel *A*. *Bull* in Semitic is *alpu* (Ugaritic) or *elef* (Hebrew), which became the meaningless word *alpha* in Greek.

The derivation becomes clearer if you take an uppercase letter *A*, turn it upside down, and add a couple of eyes. Voilà! There's your original bull's head from which our letter *A* is derived.

65

RUNES IN THE RUINS

Viking graffiti

When diggers—calling them archaeologists would be too kind—broke into the 5,000-year-old Maeshowe passage grave on the Scottish Orkney Isles in 1851, they soon realized they weren't the first intruders. The walls of the Neolithic tomb were festooned with about 30 Nordic runes dating from the twelfth century, one of the richest collections of such writing ever found. And, in a lovely example of serendipity, we know not only who carved the graffiti but also when they did so: an entry in the Norse *Orkneyinga* saga, written nearly a century later, relates how a group of Viking warriors led by Earl Harold sheltered there during a fierce snowstorm in AD 1153.

The messages on the stones range from the "Kilroy was here" variety ("Haermund Hardaxe carved these runes," "Benedikt made this cross") to your standard public toilet scratchings ("Thorni fucked, Helgi carved," "Ingigerth is the most beautiful of all women" next to a crude image of a slobbering dog). The grave site, buried beneath a cairn, isn't that large. No more than 30 people could fit in it. When a dozen of us visited a few years ago, we felt quite cramped. Yet all the runic messages appear to be carved by different individuals, meaning that most—or all—of these rough-and-tumble Viking war dogs were literate.

To the untrained eye, these crudely carved notes are illegible: long, spindly, vertical strokes with diagonal twiggy lines that distinguish one

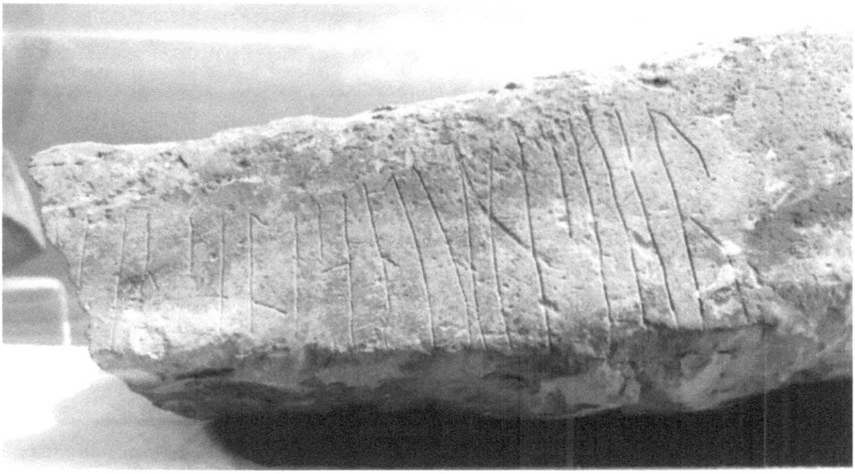

"Are in heaven, hallowed be . . ." reads this runic fragment from the Lord's Prayer on a stone found near Ophir, Orkney. It's now in the Kirkwall Museum. AUTHOR

letter from another. In fact, they're easily relatable to our own alphabet. No surprise there since both scripts have common roots dating back to Old Italic, a family of writing systems, including Etruscan, used in what is now Italy in the early days of the Roman Empire. The earliest variety of runic script, "Elder Futhark," dates back to the first century AD. (The word *Futhark* comes from the first six letters of the runic alphabet: *F*, *U*, *TH*, *A*, *R*, and *K*.) A comparison of some of the letters makes the connection with our own alphabet clear, seen in the accompanying graphic.

Other than later examples written on parchment, one obvious characteristic of all versions of runes is their *angularity*. Like most old alphabets, they were designed for carving in stone or, more importantly, wood, absent any curved lines or horizontal strokes, which, being parallel to the grain, could split the wood.

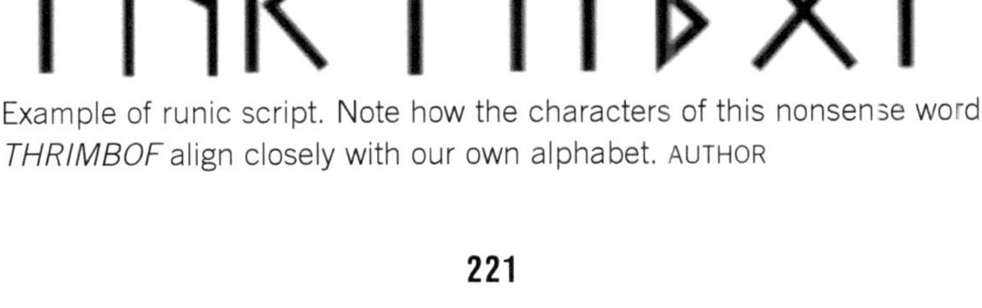

Example of runic script. Note how the characters of this nonsense word *THRIMBOF* align closely with our own alphabet. AUTHOR

How did Roman-style script end up in northern Europe two millennia ago? Probably via German mercenaries. Between the first century BC and the fifth century AD, Romans increasingly hired Germanic warriors to do their fighting for them, and it's easy to imagine Roman customs and skills, including writing, being adopted by less civilized (in Roman eyes) tribes north of the Rhine.

Runes haven't quite died out. Nazi "scholars" adopted them in their attempt to link Nazi Aryan mythology with Nordic mysticism, and you can still buy books on runic divination at any bookstore, not to mention *The Hobbit*, J. R. R. Tolkien's original "Middle Earth" epic, in which cryptic runic symbols appears on Thrór's map.

Actually, not so cryptic, with a little help from Google.

66

BLACK ENGLISH

Simplify, simplify

Black English. . . is a kind of tune-up of the English language. Some of the needless complexities are wiped away, just as happened all along the journey from Proto-Indo-European to English.

—LINGUIST JOHN MCWHORTER

Thanks to rap music and countless TV shows and movies, we're all familiar with "Black English," what civil rights leader Jesse Jackson (1941–2026) famously called "slang" and "garbage." The civil rights leader was railing against the 1996 decision by the Oakland City Council to incorporate into the classroom the dialect spoken by most black kids, so-called Ebonics, from *ebony* + *phonics*.

Ain't nothin' wrong with Black English, and IMHO, Jackson was out of line. It's not, as linguist Geoffrey Pullum patiently explained, "standard English with mistakes." Black English (formally, African American Vernacular English) is a dialect of English, no less complex, systematic, and nuanced than that spoken by National Public Radio newsreaders.

In some ways, it's more expressive than Standard American English (SAE), being "the variety of the English language that's generally used in professional communication in the United States and taught in American schools."

Black English has several grammatical features that distinguish it from SAE, including the following:

- Missing copulas: A copula is a useless bit of linguistic baggage in English that Black English has, rather sensibly, dropped: "How you doing?" instead of "How are you doing?" The copula *are* isn't the usual use of the verb *to be*, as in "My name is Barry," where "Barry" and "my name" refer to the same thing.
- Double negatives: "I ain't never seen such a sight." This is negative concord, emphasis by repetition. (Compare the less expressive "I haven't ever seen such a sight!") It's used the world over (*¡No necesito nada!*); Chaucer and Shakespeare regularly used double, even triple, negatives. For example, in *Love's Labour Lost*, Holofernes expresses his frustration: "Via, goodman Dull! thou hast spoken no word all this while." Dull replies, "Nor understand none neither, sir."
- Uninflected verbs: "She write her memoir." "He work for the county." That is, there's no *-s* ending in the present-tense, third-person singular.
- Habitual aspect: "He be singing" means he sings regularly, as a hobby or profession; that is, he's a singer.

So while Black English is often more expressive than SAE, it's generally simpler. That's because it originated as an "exoteric" language, one learned imperfectly by adults, as opposed to an "esoteric" native language learned from birth. The adults in this case were, of course, slaves transported from Africa who learned English from their white overseers. Like any exoteric language, its grammar is simplified. (It's a myth, by the way, that Black English has its grammatical roots in the dozen or so West African languages spoken natively by slaves.)

But the language spoken by those slave owners—English of the 1600s and 1700s—was originally exoteric too, as were its predecessors. Middle English, for example, derived from Old English (with all its Proto-Germanic complexity) spoken in England following settlement by

Proto-Indo-European (4500-2500 BC)

Proto-Germanc

West Germanic

North Sea Germanic (Invaeonic)

Anglo-Frisian

Old English

Middle English

Modern English

Black English (1600s - present)

One way of appreciating "Black English" is to see it as a simplification, a trend going back thousands of years, reducing the complexity of the language. AUTHOR

Angles, Saxons, Jutes, and Frisians. The next wave of invaders, Scandinavian Vikings, were mostly males who, marrying native women, learned and simplified the local language, passing this "Old English lite" (that is, Middle English) to their children. In the process, English lost most of its noun genders, verb conjugations, and much else.

This process of language simplification goes back even further. Proto-Indo-European (PIE) was an unholy mess of grammar and syntax before Proto-Germanic simplified it (see "The Phoenician Connection," in which I discuss how Proto-Germanic probably originated as an exoteric language when Phoenician adults encountered PIE).

In each subsequent iteration, it lost some complexity, such as when Old English morphed into Middle English, as we saw above. In the broader context, then, the Black English dialect is carrying on a venerable tradition of simplifying, in this case, stripping SAE of some of its complex forms.

A process surely to be continued, 'cuz you ain't seen nuttin' yet.

67

WHENCE "BRITAIN"?

Not Brutus

All the Britons dye themselves with woad which produces a blue color, and makes their appearance in battle more terrible.

—JULIUS CAESAR, *DE BELLO GALLICO*

The origins of the names of most countries are mostly noncontroversial. For example, *America* derives from the Italian navigator Amerigo Vespucci (1454–1512), who made two trips to the New World and subsequently claimed to have discovered it. Cartographer Martin Waldseemüller apparently believed him, publishing a map in 1507 with the name "America" on what is now called South America. *Canada*, meanwhile, comes from the Huron-Iroquois word *kanata* ("village"), popularized by French explorer Jacques Cartier. And *France* was the realm of the Franks, which probably meant "free people."

How about *Britain*? This is a confusing word at best since we have Great Britain (the island comprising England, Scotland, and Wales) and the British Isles (ditto, plus Ireland, the Isle of Man, and some 6,000 smaller islands), while *Britain* on its own is obsolete.

In the Middle Ages, everyone knew that the name "Britain" derived from Brutus of Troy, the guy who landed on the coast of the county of Devon at a time when giants inhabited the island. Since the giants had no weapons, they were easy prey to arrows fired by Brutus's men. When only their leader Goemagot was left, he was challenged to a wrestling match by the warrior Corineus, who promptly tossed him over a cliff. As proof, you can still see his bloodstains in the red rock below Salcombe Castle, Devon. The twelfth-century cleric Geoffrey of Monmouth penned this gory tale in his *History of the Kings of Britain*, the same book in which King Arthur is legitimized as a noble descendant of Brutus.

Brutus wasn't Geoffrey's invention, however. We can trace that back another 300 years to the *Historia Brittonum*, an anonymous ninth-century "historical" compilation in which Brutus is identified as the grandson of the Trojan hero Aeneas. According to the Roman writer Virgil (70–19 BC) in his monumental poem *Aeneid*, Aeneas, son of Prince Anchises and the goddess Aphrodite/Venus, was one of the few Trojans to escape when Troy fell. After many adventures worthy of epic poetry, Aeneas ended up in Italy, where his son Ascanius founded the precursor of the city of Rome. In Geoffrey's account, Brutus was exiled from Italy after accidentally killing his father in a hunting accident. He wandered all over, having the requisite passel of adventures before landing on "Totonesium litus," the seacoast at Totnes in Devon, hence the Brutus Stone you can still see on Totnes's Fore Street. Brutus, having given his own name to the island (and Corineus's name to Cornwall), went on to found the city of Troia Nova (New Troy) on the banks of the River Thames. The name corrupted to "Trinovantum" before it became Roman "Londonium," or "London."

So much for legends. Trinovantum actually derives from the Trinovantes people of Iron Age Britain. As for Brutus, he seems to have been a much later invention than Virgil's *Aeneid*. The name "Britain" almost certainly derives from the Celtic word *Pretani*, meaning "painted ones" (per Caesar), surviving in the modern Welsh name for Great Britain, "Prydain." The Greek explorer Pytheas of Massalia, voyaging to northwestern Europe around 325 BC, may have been the first to link Pretani with the Britannic islands.

I think I'll stick to "United Kingdom."

68

THE ROSETTA STONE

Deciphering the hieroglyphs

When Napoleon set his sights on clearing the British out of Egypt in 1798, he took with him 160 savants in his 400-ship invasion fleet. These worthies were scientists, artists, and historians who, supposedly, would bring to light the lost civilization of the Egyptian pharaohs, thus ensuring Napoleon's legacy not only as a conqueror but also as a scholar and a gentleman—which is how the so-called Rosetta Stone came to be preserved. When, in July 1799, workers rebuilding a fort at present-day Rashid (near Alexandria) happened on a heavy chunk of granite-like granodiorite, the savant-on-the-spot, Pierre-François Bouchard, saw its inscribed polished face and, presciently, realized it might be of value to linguists and philologists.

Two years later, the 1,700-pound stone was installed in a custom-built wing at the British Museum in London. Why not Paris? Because Nelson destroyed the French fleet off Alexandria on August 1, 1798, and the stone was subsequently claimed by the Brits as a spoil of war (the Egyptians not being consulted). Why the fuss over a four-foot-high slab of stone?

Because even a cursory glance shows that it's inscribed in not one but three writing styles: *hieroglyphics* (the "lost" Egyptian priestly writing, not used since the Roman emperor Theodosius outlawed pagan worship at the end of the fourth century AD), *demotic* (a "shorthand" version of hieroglyphic script, also lost), and *Ancient Greek* (well known and understood, then and now).

Once installed in the British Museum, copies—both plaster and "rubbings"—were sent to centers of learning in Europe and to Philadelphia, where amateur and professional philologists gave it their best shot at matching the unknown hieroglyphic and demotic writing with the Greek. (They knew the texts did match because the Greek text said so.)

The most visited object in London's British Museum. AUTHOR

It took another 20 years to figure it out at which time—suddenly—the thousands of hieroglyphs on the walls and columns of ancient Egyptian temples and tombs were readable, shining a light on a great civilization previously known only from its architectural wonders.

YOUNG AND CHAMPOLLION

Two decipherers did the heavy lifting. Thomas Young was an English physician and gentleman-scientist, one of several polymaths dubbed "the last man who knew everything." Young's initial breakthrough was to realize that the stone's "cartouches," oval-shaped enclosures in the text naming non-Egyptian rulers, were phonetic, starting with the characters *p t w l m y s*, that is, *Ptolemaios*, Greek for "Ptolemy."

Frenchman Jean-François Champollion, 17 years Young's junior, was a brilliant linguist who had learned, among several other languages, Coptic (still used by the Egyptian Orthodox Church), which has its roots in Ancient Egyptian. Champollion ran with Young's cartouche interpretations, but that turned out to be just the start in deciphering the complex hieroglyphics. By 1822, after five years of intense work, he had constructed an alphabet of phonetic hieroglyphic characters. From there, it was downhill. He subsequently wrote that the script was "at the same time figurative, symbolic and phonetic, in one and the same text [and even] in the same word." No wonder it took so long! Both Young and Champollion died early, at 55 and 41, respectively, honored scientists in several fields.

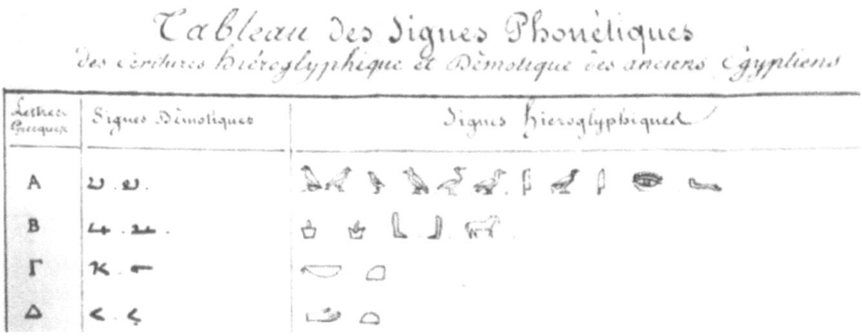

Part of Champollion's interpretation of hieroglyphs in an 1822 letter. PUBLIC DOMAIN

The text of the Rosetta Stone itself, which reappeared in at least three other stelae subsequently found, is rather uninteresting. It comprises about two-thirds (the rest is broken off and lost) of a stela dated to 196 BC and erected on behalf of the newly crowned pharaoh, seven-year-old Ptolemy V, whose great-great-grandfather had been one of Alexander the Great's generals, and who had inherited Egypt after Alexander's death in 323 BC.

Since the Ptolemaic dynasty was Macedonian, the official language of the day was Greek, hence the Greek text on the stela. However, Ptolemy V came to the throne in a period of turbulence in Egypt, a time when many different political factions had to be appeased. The most powerful one was the priestly clan of Memphis, so the stela—which essentially confirms the divinity of Ptolemy V—was diplomatically inscribed in the ritual "language of the gods" (hieroglyphic) as well as everyday Egyptian (demotic).

It wasn't the text itself but rather the opportunity that the stone gave to linguists and hence to historians that makes the Rosetta Stone so valuable. Today, it's the most visited object in the vast British Museum collection and one of the most controversial. Why shouldn't it be returned to the country of its origin? Egyptian archaeologists want to know. Like the "Elgin Marbles" (saved/stolen from Greece in the early 1800s), the story of the Rosetta Stone is unfinished.

69

WHEN DID LANGUAGE BEGIN?

100,000 or a million years ago

Seriously! The title isn't hyperbole; it's what linguists argue about. This isn't like the age of the universe, 13.7 billion or 13.8 billion years old, which is the same thing for most of us. This is a serious, 10-fold difference of opinion. If you're in the "continuity" camp, you believe that language arose gradually a million or more years ago in *Homo erectus* ("upright man"), the species of humans that led to *Homo neanderthalensis* and us, *Homo sapiens*, among others. The "discontinuity" camp, led by the redoubtable linguist Noam Chomsky, thinks that language came about as a result of a single chance neural mutation in one member of the newly emerged *Homo sapiens* between 100,000 and 80,000 years ago.

Unlike tools and bones, language leaves no trace. Everything we can say about its origin has to be inferred from indirect evidence, with many scientific disciplines adding to the picture: linguistics, archaeology, comparative biology, cognitive science, paleoneuroscience, evolutionary theory, genetics, and more. The current consensus has swung away from Chomsky (who practically ruled the field of linguistics for decades) and his followers. The proponents of the gradualist camp say that a single individual, born with the capacity for language as a result of a mutation, wouldn't have had an advantage over his or her nonlinguist peers. I'm reminded of H. G. Wells's short story *The Country of the Blind*, where the only sighted person is an unwelcome—and vulnerable—aberration.

(There was a brief flurry of excitement in 1998, when a gene, FOXP2, was shown to be critical in the production of language. Was this Chomsky's hypothesized "language gene" that allowed our ancestors to go from nonspeaking to speaking in a few generations as it swept through ancient populations? No. It turns out that language is far more complicated than can be explained by a single mutation, and other genes are also necessary.)

More compelling for most researchers is a scenario in which language—defined as "transmission of information by symbols"—arose *gradually*, starting at least a million years ago. This was a time when the smartest creature to have ever lived, *Homo erectus*, was fashioning bone and stone tools, taming fire for cooking, creating art, and navigating across oceans to the islands of Flores (Indonesia) and Socotra (off the Horn of Africa). That is, they possessed *culture*, the precursor of language. According to linguistic anthropologist Dan Everett, "Culture transforms 'things' into symbols and meaning. If *erectus* had symbols, it had language."

While researchers are slowly coming to a consensus about the *who* (*Homo erectus*), *when* (1 million or more years ago), and *where* (Africa) of language, that leaves the *how* and *what* unanswered. Theories for how language came about abound, with primate communication—grunts, yells, and gestures becoming increasingly complex—frequently cited. What language consisted of back then is also disputed: while the Chomsky camp maintains that grammar is essential for language, the gradualist school says that you don't actually need grammar for communication. Taking Everett's examples, everyone knows what the grammarless sign "No shirt, no shoes, no service" means, while the movie title *Eat Drink Man Woman* practically gives the whole plot away! And these days, a few simple emoji symbols can communicate whole realms of meaning sans grammar.

The debate over how "human's greatest invention" came about is far from over, but increasingly, it seems that we and our Neanderthal cousins were born into a world already rich with language.

70

THE VINDOLANDA TABLETS

Voices from Roman Britain

Oh how much I want you at my birthday party. You'll make the day so much more fun. I do hope you can make it.

—LETTER FROM CLAUDIA SEVERA TO HER SISTER SULICIA LEPIDINA, WIFE OF FLAVIUS CERIALIS, PREFECT OF A COHORT STATIONED AT VINDOLANDA FORT, NORTHERN ENGLAND, CA. AD 100

This birthday invitation written on a thin wooden tablet may be the oldest surviving writing in Latin by a woman. Thanks to being preserved intact beneath thick layers of mud, it is one of hundreds of handwritten documents found at Vindolanda Fort illuminating life on the frontier of the Roman Empire nearly 2,000 years ago. Most of these thin post-card-sized "tablets" are now in the British Museum, where they have been voted "Britain's Top Treasure."

The Romans conquered most of "Britannia" in the years following Emperor Claudius's invasion of AD 43. In the north, where Pict tribes fiercely resisted the incursion, Roman legionnaires built a series of forts, first out of wood and, later, stone. One of the most important of these was Vindolanda, established around AD 85 and sited midway between present-day Carlisle and Newcastle in Northumbria. Forty years later, Emperor Hadrian ordered the construction of a 72-mile-long wall across

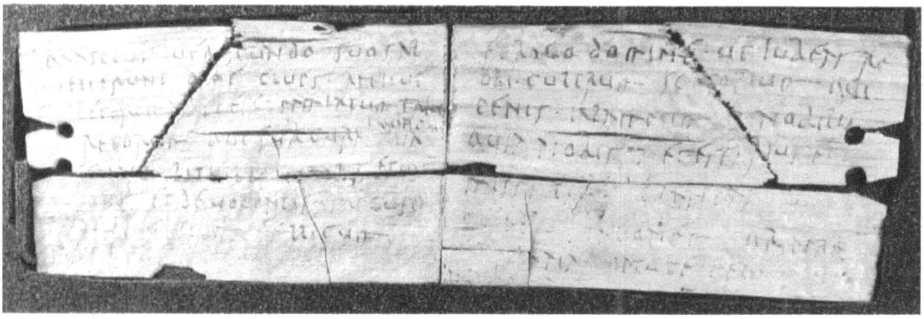

This two- by six-inch wooden tablet in the Vindolanda museum begins "Masclus Verecundo suo sal(utem) . . ." ("Masclus to his Verecundus, greetings . . ."). He goes on to request leave for several of his men. AUTHOR

the width of the country as a defense against the Picts and to control trade between local tribes. "Hadrian's Wall," originally 16 feet high (now much reduced after centuries of pillaging the stone blocks for local buildings), runs a mile north of Vindolanda Fort.

The first of the Vindolanda tablets was unearthed during an archaeological dig in 1973. Thought at first to be two wood shavings stuck together, when peeled apart, they revealed writing on the inside that quickly faded in the air. However, infrared photography made the ancient script visible again. Since then, nearly 1,000 similar tablets have been found and deciphered.

The messages, written in cursive Latin script on thin slivers of birch or alder, reflect day-to-day concerns: *military matters* (Can the surveyor Crispus please have a lighter military service?), *requests for provisions* (My fellow soldiers have no beer . . . please order some to be sent), *commercial demands* (Make sure you send me some cash so that I may have ears of grain on the threshing room floor), and *personal notes* (I have sent you ?? pairs of socks, two pairs of sandals, and two pairs of underpants).

The British Museum loaned nine of the tablets back to the Vindolanda museum, where curious visitors like myself can view them in the "vault" and be transported back to a time when the site bustled with hundreds of garrison troops from all over the empire, along with their commanders (some of whom brought their families), local farmers, and shopkeepers. The fort and its adjacent village thrived for hundreds of years, only to be deserted sometime after the last of the Roman legions left Britain in AD 410.

Starting with discoveries by amateur archaeologists in the nineteenth century, Vindolanda has come to life again, especially with the discovery of these tablets, each of which gives us a glimpse into life on the frontier of Imperium Romanum, the Roman Empire. The timeless quality of these messages—a mother worrying about her son staying warm, a commander ordering beer for his men, Claudia's birthday invite—cuts through time, making their world ours.

71

SILENT READING AND WORD BREAKS

How they're related

When [Ambrose] read, his eyes scanned the page and his voice was silent and his tongue was still.

—St. Augustine, *Confessions*, AD 397

The fact that a scholar like Augustine would express surprise on finding Ambrose, Bishop of Milan, reading *silently* tells us something about how readers approached books in the ancient world. For thousands of years, from the invention of writing in the fourth century BC to around AD 1000, reading almost invariably meant *reading aloud*. For most of that time, it also meant reading from unbroken lines of text.

Notice, for example, that the writing in the *Codex Sinaiticus*—a copy of the Greek Bible made around AD 350—is in *scriptura continua* (continuous script), with no spaces between words. For almost all readers, understanding came by sounding out the syllables: early writing was meant to be read aloud. When we talk, we—don't—speak—like—this, wespeaklikethis, and the ancients wrote the way they spoke. Stringing together letters without spaces between words wasn't a problem back then.

Around AD 800, Irish monks invented the word-break system, leading to near-universal silent reading since spaces between words relieved readers of the need to sound out text. Actually, the monks *re*invented word breaks. Previous to 800 BC, when ancient Greeks adopted the

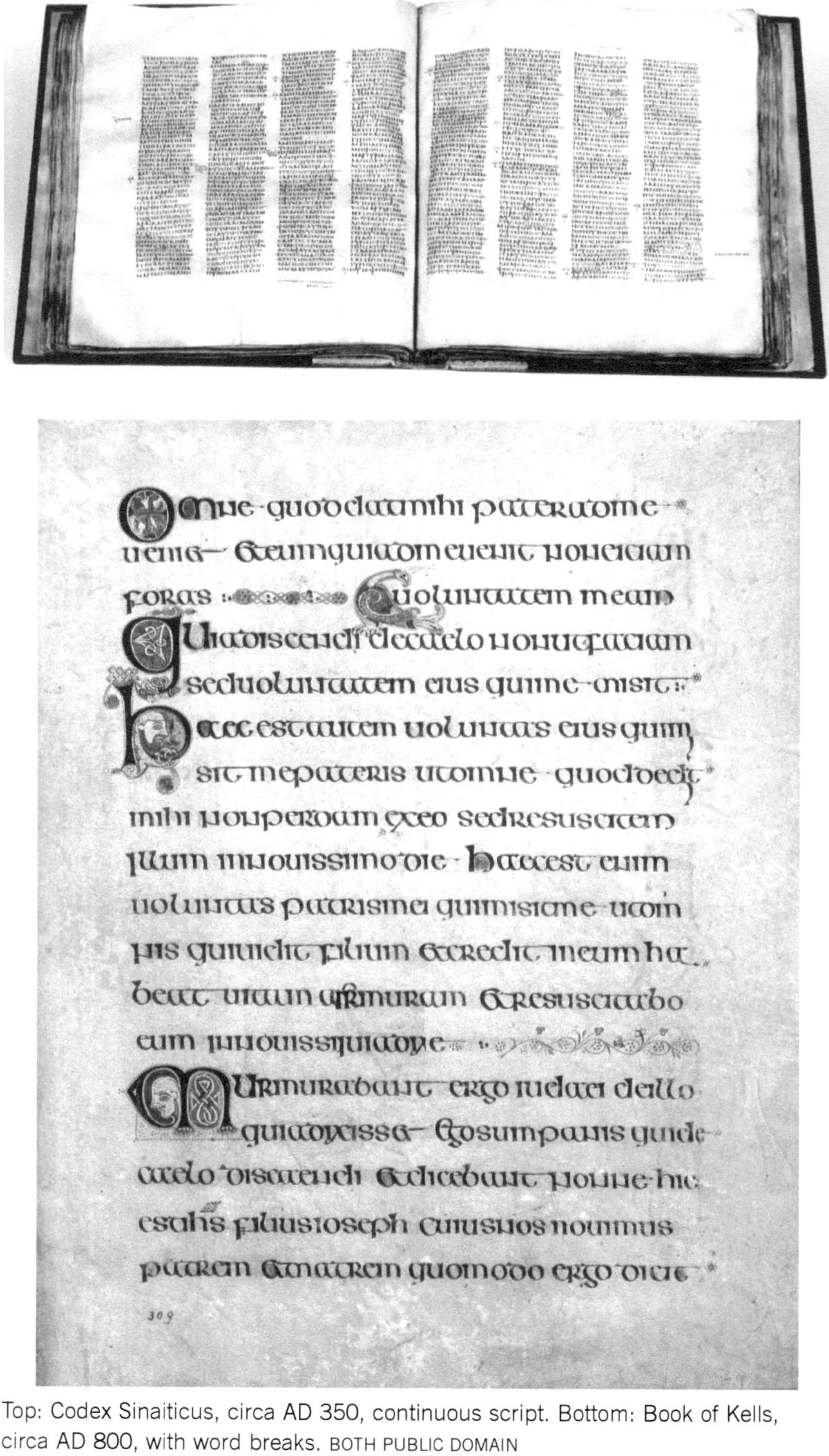

Top: Codex Sinaiticus, circa AD 350, continuous script. Bottom: Book of Kells, circa AD 800, with word breaks. BOTH PUBLIC DOMAIN

Phoenician alphabet, most writing in the ancient world *did* have spaces between words, but those words lacked vowels. When Greek scribes added symbols for vowels, they apparently considered word breaks an unnecessary convenience, and for more than 1,000 years, Greeks, then Romans, wrote almost exclusively in *scriptura continua*.

Why would those intelligent, literary people drop what seems to us such an obvious advantage for the reader? According to literary detective Paul Saenger in *Space between Words* (in which he devotes nearly 500 pages to the question!), "[Readers] relished the mellifluous metrical and accentual patterns of pronounced [that is, spoken] text and were not interested in the swift intrusive consultation of books [so] the absence of interword space in Greek and Latin was not perceived to be an impediment to effective reading."

When scribes began to add spaces between words, it changed the whole neuropsychological process of reading. No longer did a solitary reader have to slowly and painfully extract information from text, mumbling to him- or herself as they did so. With the task of word separation now assigned to the scribe rather than the reader, reading became silent, rapid, effortless, and meditative, and what had formerly been a chore became a delight.

Since you've gotten this far, I trust you agree.

72

FOR THE WEEK-HEARTED

Our curious seven-day period

Every seven days since March 3, 1923, a new edition of *Time* magazine has appeared, regular as clockwork. What's so special about that number seven that it forms the backbone of our calendar? After all, the week follows no natural cycle, as is obvious when you realize that other cultures have gotten along quite happily with three-day weeks (ancient Colombia), four (much of Africa), eight (early Rome), 10 (ancient Peru and revolutionary France), and even 19 (Baha'i faith).

When we agree to call the first day of the working week Monday ("moon-day"), we're keeping a particular faith, dating back to at least 100 BC, that maintains that every seventh day is ruled by the moon. Our ancestors have been acknowledging every seventh day to be the moon's for more than 2,000 years. Like an endless chain letter (despite adjustments to the calendar), the cycle of days of the week has been maintained faithfully in the West by the Greeks, Romans, Angles, Saxons, Vikings, and all the rest who came before us. We just happen to be the latest generation to keep the faith.

Monday/moon-day might give you a clue as to why we have seven days in our week. The moon is a bright astronomical object, as is Saturn (Saturday) and the sun (Sunday). That takes care of three of the seven moving sky objects known to the ancients. The other four days correlate to the rest of the bright "wanderers" (Greek *planētai*, "planets"), whose names in English come from Norse mythology under the influence of

The Babylonian *Imago Mundi*, dated to the sixth century BC, shows Babylon on the Euphrates surrounded by a circular landmass with a seven-pointed star corresponding to the seven heavenly "wanderers." BRITISH MUSEUM, PUBLIC DOMAIN

Angles, Saxons, and Jutes, all of whom settled England (Angle-land) in the so-called Dark Ages. Because most of the Nordic gods were associated with Roman gods during the heyday of the Roman Empire, we can trace those names back to Roman and Greek sources.

Tuesday: Old Norse *Tiw*, the Teutonic god of war, was identified with Roman *Mars*. The Old Norse name is cognate with Latin *deus* ("god"), which in turn comes from Greek *Zeus*, "father of gods and men." (To complicate matters, *Zeus* was identified with *Jupiter* in the Roman pantheon, but he gets his own day, Thursday.)

Wednesday: The day of *Wodan*, or *Odin*, was connected to *Mercury* by the Roman historian Tacitus.

Thursday: *Thor*, eldest son of *Odin*, was the Scandinavian *Vulcan*, god of thunder. He's associated with the Roman planet and god *Jupiter*.

Friday: From *Freya* or *Frigga* (curiously cognate with "free"), Scandinavian goddess of love and wife of *Odin*. As the female member of the pantheon, she's partnered with Roman *Venus*, who in turn is connected with Greek *Aphrodite*.

Seven bright wanderers, seven days in a week: it's written in the stars.

73

HESIOD

Farmer, poet, misogynist

[Hesiod's] personality behind the poems is unsuited to the kind of "aristocratic withdrawal" typical of a rhapsode but is instead "argumentative, suspicious, ironically humorous, frugal, fond of proverbs, wary of women."

—The Oxford History of the Classical World

Misogyny—hatred of or contempt for women—goes back a long way, at least as far as the ancient Greek poet Hesiod, around 700 BC. (I argue in "Agriculture: Our Worst Mistake?" that the problem started with the agricultural revolution.) Three hundred years later, in the *Timaeus*, Socrates warns men that if they live immorally, they will be reincarnated as women (horrors!); later yet, St. Paul advised, "Let your women keep silence in the churches: for it is not permitted unto them to speak, but they are commanded to be under obedience." More recently, Friedrich Nietzsche claimed that "stricter controls on women was a condition of every elevation of culture." Think we're beyond all that? Despite passage by both houses of the U.S. Congress more than 50 years ago, the Equal Rights Amendment ("equality of rights under the law shall not be denied or abridged . . . on account of sex") has never been formally ratified.

Hesiod, who probably lived in Boeotia in central Greece at about the same time as Homer, was a farmer given the gift of poetry by the Muses. Most authorities believe he was a real person and that he obtained his material both from early Greek *rhapsodes* (professional storytellers) and from Egyptian sources. He's best known for two long poems: *Works and Days*, in which he lays out the "Five Ages of Man," and *Theogony*, the origin story of the gods.

Before the creation of women, in Hesiod's telling, men experienced a Golden Age. "They lived like gods without sorrow of heart, remote and free from toil and grief. Miserable age rested not on them, but with legs and arms never failing they made merry with feasting. . . . They dwelt in ease and peace." That all came to an end when the Titan Prometheus gave men the gift of fire, having stolen it from the gods. In retaliation, Zeus (a) bound Prometheus to a rock where every day an eagle ate his liver, which grew back overnight, and (b) punished men by giving them "an evil thing for their delight": woman.

The first woman was Pandora, whom the god Hephaestus molded from clay according to Zeus's specifications. She carried a jar, Greek *pithos* (not a box), having been told never to open it. Well, duh. She did of course (otherwise, it wouldn't be much of a story), releasing into the world all the evils: labor, old age, sickness, death, worry, crime, hate, envy, and the rest. She quickly put the lid back on, preventing the last evil, *hope*, from escaping.

Why hope? The idea is that we prolong life's torment by hoping for a better life, which is always beyond our reach. Pandora was only the start of troubles for the bros. From Pandora, according to Hesiod, emerged "the race of women and female kind. Of her is the deadly race and tribe of women who live amongst mortal men to their great trouble." The petulant old poet did allow that (rarely) a man can find a good wife, but he couldn't stop himself, adding, "evil contends with good."

So much for the Golden Age. Hesiod then told of the Silver Age, when people lived most of their lives as children (and when males were dominated by their mothers); followed by a Bronze Age, during which men were tough and war was their passion so that they eventually destroyed each other; then an Age of Heroes, corresponding to the Mycenaean age, when men fought and died at Troy (because of, yup, a woman, Helen, the fairest of all); and finally the Iron Age, the gloomy age in which Hesiod lived: "Hateful strife bore painful Toil, Neglect, Starvation, and tearful Pain, Battles, Combats."

That was 2,700 years ago. Of course, things are much better now.

MYTHS

The stuff of dreams

74

THE VOYNICH MANUSCRIPT

It's been decoded . . . again!

The mysterious Voynich manuscript has finally been decoded.
—*Ars Technica* headline, September 8, 2017

Every two or three years, ever since the so-called Voynich Manuscript resurfaced from obscurity in 1912, professional researchers and amateur decipherers announce they've decoded this fifteenth-century book. In 1917, Greg Kondrak, professor of computer science at the University of Alberta, used artificial intelligence techniques to translate not the entire book, despite the above headline, only the first sentence! Here it is: *She made recommendations to the priest, man of the house and me and people.* "It's a kind of strange sentence to start a manuscript," he admits, "but it definitely makes sense." (If you say so.) A year later, the journal *Digital Philology* rejected a submission from Ahmet Ardıç, a Turkish Canadian electrical engineer, who claimed that the manuscript was written in Old Turkic. For a recent attempt at decoding, check out the September 2024 issue of *The Atlantic*. By the time this book is published, you'll be able to read at least a couple more interpretations.

Kondrak and others join a long list of would-be decoders, starting in 1921 with a University of Pennsylvania researcher who ignored the actual visible text; instead, he proposed that the real message was in "Greek shorthand" buried within individual letters and visible only through a microscope. (No one else could see them.) Since then, claims for

Detail from one of the Voynich Manuscript's 200-odd pages, written with a quill pen on 600-year-old calfskin. PUBLIC DOMAIN

a Greek/Latin/Egyptian/Ukrainian/Arabic/Hebrew/Flemish/English basis for the manuscript have regularly appeared—and shown to be false.

The manuscript itself is a vellum calfskin codex (bound book) of 240 pages, about 38,000 words, out of an estimated 272 original pages. The skin has been radiocarbon-dated to between 1404 and 1438, thus establishing the earliest date for the writing. Although attempts at deciphering have so far failed, the colored illustrations tell us that the book includes sections on herbs and botany, astronomy and cosmology, and pharmaceutical cures. And nymphs, lots of nymphs.

Whatever the curious symbols mean, they appear to be letters making up an alphabet of 25 (more or less) characters. Words are between 2 and 10 letters long, with no punctuation. Because it's written in an unknown language in an unknown script, it's generally assumed to be in code. However, there's no indication of any delay or hesitation between characters, which would be expected if the scribe were encoding while writing. And, even more curious, the unknown author corrected not a single character out of the total 170,000 letters in the manuscript! The

MR. VOYNICH AMONG HIS BOOKS IN SOHO SQUARE

Wilfrid Voynich, 1899. UNKNOWN PHOTOGRAPHER, PUBLIC DOMAIN

fluid nature of the script seems to be telling us that the book's architect wrote in a natural script in their natural language.

Skeptical readers will surely be asking by now if the whole thing is a hoax, either created several hundred years ago as a moneymaking scam (Emperor Rudolf II, 1552–1612, supposedly paid 600 ducats—73 ounces of gold—for it) or a modern forgery. If the latter, the prime suspect is the Polish book dealer who gave his name to the book: Wilfrid Voynich, who said he bought it in 1912. However, this seems far-fetched in the light of the radiocarbon dating. How could he have acquired so much untouched 500-year-old calfskin?

H. P. Krause, the dealer who donated the manuscript to Yale, likened it to the Sphinx: "its lair strewn with the bones of those who failed to solve the riddle, and still awaiting the Oedipus who will give the right answer." (Good luck with that!)

The mystery remains, having stumped experts in medieval history, code breakers from the two world wars, professional linguists, and amateurs from all over. It's available online at https://archive.org/details/TheVoynichManuscript.

I'm still unconvinced, but let me know if you figure it out.

75

DOUBTING SHAKESPEARE

Stratford man or Stratford myth?

They're both bonkers!

—Stanley Wells, arguably Britain's leading Shakespeare expert, referring to Shakespearean actors Derek Jacobi and Mark Rylance, who are convinced that the Stratford man didn't write the Shakespeare canon

On April 26, 1564, the baptism of one "William Shakspere" was recorded in the provincial town of Stratford-upon-Avon in Warwickshire, England. According to the conventional story, this glover's son grew up to become the greatest writer in the English language, author of 38 plays, 154 sonnets, and three long poems. But was Stratford's "Shakspere" *the* "Shakespeare" (or hyphenated "Shakes-peare") whose name appears on the title pages of these works? Or, as seems increasingly likely, was the Shakespeare canon written by someone else while Shakspere, wool merchant and sometime actor, had nothing to do with it? Like many before me, I believe that the Stratfordian case—that the Stratford man and the writer were one and the same—is so flawed as to be unbelievable.

ANTI-STRATFORDIANS

The anti-Stratfordian position started out as a series of nudge-nudge-wink-wink hints in the early seventeenth century, which implied that the name on the printed title pages was the pseudonym of someone who needed to remain anonymous. The case has only improved over the years. Despite this, orthodox biographies of Shakespeare are appearing about once a year, perhaps because writers who identify the Stratford man with the great writer have it easy. With so few hard facts about his life, they can make up any tale they want. And they do! For example, Stephen Greenblatt's 2004 bestseller *Will in the World* (which begins, "Let us imagine . . ." and continues in the same vein) was downplayed by *The Times* reviewer as "biographical fiction."

Anti-Stratfordians claim that the immortal Shakespeare whom we know and love was probably not the tax-dodging businessman from a provincial town who could barely sign his own name; or the father who made no effort to educate his daughters regardless of many intelligent and well-read female characters in the plays; or the 52-year-old man who died unmourned and unnoticed despite—supposedly—being the most famous playwright of his time; or the puffy-faced guy whose bust in his local parish church originally showed him as a wool merchant (his wool sack was replaced years later with a writing surface, and he was given a quill pen!); or the cultured genius who was able to read and write in French, Italian, and Greek and who possessed an intimate knowledge of northern Italy and France.

I say "probably" since, as far as the majority of Shakespearean scholars are concerned, all the above can be explained away, and anyone who doubts that the Stratford man couldn't have penned the works attributed to William Shakespeare must be ill-informed, deluded, or, as one scholar had it, immoral! Shakespeare is, after all, an institution and "bardolatry" a thriving business, perhaps England's greatest export.

As such, it's the saga of a young man coming from a poor, illiterate family who mysteriously morphed into the writer of works "dazzling in their erudition, steeped in the learning of Renaissance humanism," quoting Elizabeth Winkler's *Shakespeare Was a Woman and Other Heresies*. This rags-to-riches story wasn't unprecedented. Christopher Marlowe came from similar circumstances, but, unlike Shakspere, we know that his genius was recognized early and that he attended Cambridge University on a scholarship. You can be born a genius, but knowledge has to be acquired.

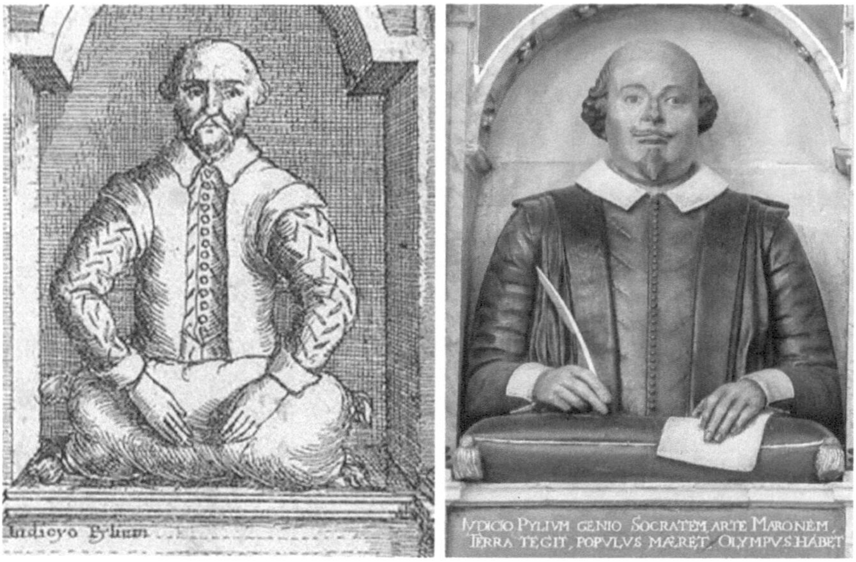

Before-and-after images of the Shakespeare funerary monument in Holy Trinity Church, Stratford-upon-Avon, which originally showed him with a sack of wool on his lap. After "restoration" in 1748, he was given a pen, and the sack morphed into a writing surface. No mention is made of him as a great writer in the ambiguous text below. Before: Dugdale's 1656 *Antiquities of Warwickshire*, engraved by Hollar. After: Tom Reedy. BOTH PUBLIC DOMAIN

THE FLAWS

Here are some of the flaws in the tradition that the man from Stratford wrote the Shakespeare canon:

- Signatures: We have just six examples of the Stratford man's signature. Could someone who could barely write his own name have penned nearly a million words?
- The will: The Stratford man's will (in which he famously bequeathed his second-best bed to his wife), while giving detailed instructions for the distribution of his assets, makes no mention of unpublished plays, poems, books, manuscripts, or maps, nothing that might be expected in the will of a prolific writer steeped in the knowledge of the day.

The only known signatures of the man purported to be the greatest writer in the English language. PUBLIC DOMAIN

- Erudition: Shakespeare the author really knew his stuff! The plays reference books in French and Italian that hadn't been translated into English at the time, he wrote accurately about cities in northern Italy and knew contemporary French and Italian colloquialisms (three scenes of *Henry V* are entirely in French), his descriptions of upper-class life mirror someone intimately familiar with the courts of Elizabeth I and James I, and he knew the law (at least 50 books have been written discussing his legal proficiency), not to mention his knowledge of astronomy, medicine, warfare, mythology, music, typography, geography, and much more. How could a wool merchant from a provincial town have become so erudite?

- The First Folio: This is usually offered as prima facie proof for the Stratfordian case. Published in 1623, seven years after the Stratford man's death, the "First Folio" of 36 Shakespeare plays is often cited as one of the most influential books ever printed. It refers to the author as "Sweet Swan of Avon," which, unfortunately for Stratford's true believers, isn't quite as cut-and-dried as it appears. Avon, Celtic for "river," is the name of five other waterways in England besides the Stratford-upon-Avon one. It gets worse: Hampton Court, where many of Shakespeare's plays were performed for the royal court, was referred to as Avondunum ("river castle"), or Avon for short. Nothing else in the First Folio comes close to equating the Stratford man with the playwright. In fact, it's suspiciously short on saying anything specific about the author beyond vague plaudits.
- The Stratford man's documents: Researchers have found about 70 documents referencing the Stratford Shakspere, most of which portray him in a bad light and none as a writer. He made a habit of suing his neighbors for small sums, he was listed as being delinquent in his tax payments, he was accused of hoarding grain during a famine, and he had a restraining order drawn up against him, all of which make him sound like a petty (and rather obnoxious) businessman. We do have records of his twice being paid as an actor but—tellingly—not as a writer. Theater impresario Philip Henslowe, who staged Shakespeare plays, made payments to 27 playwrights but none to Shakespeare.
- Shakespeare the feminist: "Shakespeare's sympathy with, and almost uncanny understanding of, women characters is one of the distinguishing features of his comedy, as opposed to that of his contemporaries," according to Shakespeare scholar Anne Barton.

Consider such intelligent and witty characters as Portia (*The Merchant of Venice*), Rosalind (*As You Like It*), Viola (*Twelfth Night*), Beatrice (*Much Ado About Nothing*), Hermia (*A Midsummer Night's Dream*), Katherina (*The Taming of the Shrew*), and so on. Were these the creations of a man who was indifferent to the literacy of his daughters? Or of a man at all?

WHO, THEN?

If the man from Stratford didn't write the Shakespearean canon, who did?

Originally, Francis Bacon was considered the prime candidate, but his attested writing is so inferior to the stuff of the plays and sonnets that he's rarely considered today. Later contenders include Christopher Marlowe; Sir Henry Neville; Mary Sidney Herbert, Countess of Pembroke; William Stanley, 6th Earl of Derby; and many more, nearly 100 in all! Or it was an amalgam of these, writing under a single pseudonym. Today, the candidate most frequently cited is the 17th Earl of Oxford, Edward de Vere, with Neville playing catchup.

So-called Oxfordians note that de Vere had both motivation for writing under a pseudonym and the qualifications: he had a classical education (he attended Cambridge) and a law background (he studied at Grays Inn law courts), and he was worldly, having traveled extensively in Italy and France, and a confidant of Queen Elizabeth (whose chief adviser was de Vere's father-in-law, William Cecil Burghley). He was also a good writer, said to be "the most excellent" of Elizabeth's courtier poets.

De Vere's motivation for anonymity is obvious. Take the character of Polonius, chief adviser in the Danish court, portrayed as a garrulous, bumbling old man—clearly, to most Shakespeare scholars, modeled on Burghley himself. In those days, to openly mock the crown or government was considered treason, punishable by maiming, imprisonment, or death. The main strike against de Vere is that he died in 1604 so that if he was Shakespeare, several plays had to have been published posthumously, edited, perhaps, by someone in the know. Mary Sidney?

For a really wild ride, check out Ted Story's *The Shakespeare Fraud*, in which he claims that de Vere fathered a Tudor prince, the mother being

Edward de Vere in 1575, age about 25. SEVENTEENTH-CENTURY PAINTING BY AN UNKNOWN ARTIST BASED ON A LOST ORIGINAL WORK, PUBLIC DOMAIN

none other than the queen of England, Elizabeth I, which explains de Vere's reluctance to be known as the Bard of Avon.

If de Vere or someone else was posing as "Shakespeare," why that particular name? One possibility is that the Stratford man, who was involved in the London theater scene as an occasional actor and an investor in the Globe and Blackfriars Theatres, was paid to act as the face for a playwright who wished to remain anonymous. Or, more likely, Stratford's Shakspere had nothing to do with it. Shakespeare, or sometimes Shake-speare, may simply be a play on the goddess Pallas Athena, who sprang from the head of Zeus "shaking a sharp spear." William, meanwhile, is cognate with the Old Dutch for "golden helmet," which Athena is often portrayed as wearing.

Not incidentally, it was common practice in that time to indicate pseudonymous works with a hyphen, so Shake-speare would then be seen alongside such giveaways as Tom Tell-troth and Simon Smell-knave.

IN THIS CORNER, WE HAVE . . .

To legitimize the authorship debate, veteran Shakespearean actors Sirs Derek Jacobi and Mark Rylance, among others, sponsored a "Declaration of Reasonable Doubt about the Identity of William Shakespeare," which you can read at https://doubtaboutwill.org.

Jacobi and Rylance are hardly the first to question the authorship, being preceded by, among others, Mark Twain, Henry James, Walt Whitman, Ralph Waldo Emerson, Orson Welles, Sir John Gielgud, Sigmund Freud, (Supreme Court Justice) John Paul Stevens, historian David McCullough, and physicist Roger Penrose, none of whom, I trust you'll agree, were or are lacking in intelligence.

I'll leave the final word to Nobel Prize–winning novelist John Galsworthy, who declared the original proposal that Edward de Vere was Shakespeare (in a 1920 book by English schoolteacher John Looney) to be "the best detective story I've ever read." A fitting byline for the entire attempt to figure out who wrote Shakespeare is that we may never know, but it's sure fun to try!

76

VLAD, VAMPIRES, AND DRACULA

The warlord-diplomat behind the myth

Don't think it's easy being a vampire. Apart from the mundane sunlight/holy-wafer/silver-cross-in-your-face/wooden-spike-through-your-heart issues, you've got to deal with the fact that blood is toxic.

A few teaspoons won't hurt—blood pudding is a delicacy for Brits, as is seal blood for the Inuit—but start drinking pints of the stuff (like real vampires do), and you risk hemochromatosis, or iron storage disease. We all need iron, of course: ask any anemic. It's a necessary component of the hemoglobin that carries oxygen from your lungs to the rest of your body, hence the bright red color of arterial blood. However, humans have a limited tolerance to excrete excess amounts; too much iron, and your liver and nervous system go caddywonkers, unlike vampire bats, for example, whose intestines have a filtering membrane that prevents overloading their bloodstream with iron.

This toxic message hasn't, apparently, reached the daily coachloads of tourists visiting Castle Bran, aka "Dracula's Castle," in Transylvania, Romania's central region. The official brochure insists that the fourteenth-century structure has only a passing connection with Vlad Tepes III (ca. 1430–1477), supposedly the model for Bram Stoker's 1897 novel *Dracula*. But having separated Vlad from vampires, the brochure then discusses his history in the context of vampirism! Not to mention all the kitsch—fangs, vampire mugs, the whole shtick—for

Castle Bran ("Dracula's Castle"), Transylvania, in which Vlad III, the "Impaler," was briefly imprisoned. AUTHOR sale at the castle entrance. I'm surprised Stoker's heirs haven't sued the Romanian Tourist Board for royalties.

The fact is that Stoker, a Dubliner who made good in London (he managed Henry Irving, the most successful stage actor of his time), never refers to Vlad in his copious notes for the novel and mentions him in the book only in passing. In contrast to Count Dracula, Vlad was a very real person, three times *voivode* (prince) of Wallachia, popularly known as the "Impaler" after his chosen method of dispatching his enemies: slowly and painfully as they dangled on a pointed stake thrust from anus to mouth. In Romania, he's something of a folk hero, a master of psychological warfare who stopped the Turks from overrunning Wallachia and who stripped the local Hungarian *boyars* (ruling class) of their power.

Vlad signed himself "Dracula" (dragon) after the chivalric Order of Dragon founded by the king of Hungary as a mutual defense pact against the Ottoman Turks. Vlad's father had been invested into the Order, and Vlad inherited the allegiance. And no, despite popular belief, "dracula" doesn't mean "son of the devil" in Romanian. That would be "pui de drac."

So what is the connection between Vlad and Stoker's Count Dracula? According to literary scholar Elizabeth Miller, very little. "Stoker borrowed the name *Dracula* and a few scraps of miscellaneous information about Wallachian history from William Wilkerson's *An Account of Wallachia and Moldavia* (1820)," she writes. We can be glad Stoker did stumble on the name, though. Until then, his courtly villain had gone by the lispy name of "Count Wampyr."

I wonder what Stoker would make of his creation today? Featured in more than 100 movies, thousands of books, and a staple of Halloween costumes, Dracula is alive and well, immortal, in fact, which, of course, is the whole point if you're going to be one of the undead.

77

FAIRY TALES

Fooling Sherlock Holmes' creator

It seems that most of us have always believed in an afterlife of some sort, usually (but not always) pleasant. Anthropologists have yet to find a society that didn't hold beliefs about life after death, while archaeologists regularly find richly stocked graves, both human and Neanderthal, dating back tens of thousands of years. Food and drink, weapons and armor, jewelry and ornaments—all the essentials to ensure a safe passage from this realm to the next—frequently accompany skeletons. These grave goods give us a glimpse into what our ancestors believed awaits us after death.

Often the choice is binary. You'll either end up in purgatory ("Where can I get a cold beer around here?") or paradise ("Enough with the harps, already!"). And if not heaven or hell, perhaps you'll choose—or be designated—to return to this realm as a sea slug, an eagle, or (my fantasy) a fabulous ballet dancer. Maybe you'll come back as a trench-coated angel, tasked with recording the day-by-day pains and pleasures of this earthly existence per Wim Wenders' lovely movie *Wings of Desire*. Or you'll be marooned forever between worlds. Or . . .

We humans have been speculating about the Great Beyond for at least 100,000 years, when our ancestors started burying their dead, presumably on the assumption they'd be up and about at some point. Beads, rings, spears, swords, clothes, food, and drink have been found in ancient graves from India to Peru. Even Neanderthals, once written off as our

Frances Griffiths and friend in one of the 1917 "Cottingley Fairy" photos. ELSIE WRIGHT, 1901–1988, PUBLIC DOMAIN

tough but oafish cousins, often buried their dead with a few thoughtful resources for their next adventure.

THE HYGIENIC ADVANTAGE

How could near-universal belief in life after death have arisen absent any evidence? Maybe the key was *hygiene*. It could have worked like this. Two tribes living side by side are identical in every way except that some imaginative person in one tribe has a radical idea that life doesn't end at death. She or he convinces the rest of the tribe that, rather than toss their dead out of the cave to be scavenged by wild animals, they should bury them just in case the recently departed needs their intact body for what comes next.

In time, basic hygiene considerations would have come into play since rotting flesh is a prime source of disease. The bury-our-dead tribe is spared this particular threat and starts to thrive in comparison with their less healthy toss-out-our-dead neighbors. Over the millennia, the life-after-death believers flourish, and the idea eventually gets added to all the other assumptions encoded in human genes. Today, experiments show that most two-year-olds appear to have been born with a belief in life after death. (Me, I crave *simplicity*, and what could be a simpler afterlife than "nothing"?)

HEAVENS ABOVE!

You'd think after all this time that nothing more could be said about heaven or hell, but the *New York Times* nonfiction bestseller list told us otherwise a few years ago. For example, *Proof of Heaven: A Neurosurgeon's Journey into the Afterlife* topped the nonfiction (!) list for weeks. In it, an otherwise comatose neurosurgeon, Eben Alexander, describes cavorting in heaven with butterflies, angels, and, wouldn't you know, nubile chicks. Then there was *Heaven Is for Real: A Little Boy's Astounding Story of His Trip to Heaven and Back*, a toddler's tale of visiting paradise and sitting on Jesus' lap. It sold a million copies and also made the top on that same list. "We do not second-guess personal accounts," wrote Deborah Hofmann, senior editor of the *Times* bestseller lists, in defending the book's nonfiction classification.

My envy for these get-rich-quick scams is matched only by my skepticism. Yet many of the finest minds in history have been convinced in the reality of the afterlife, including our ability to communicate with the dead. Take Sir Arthur Conan Doyle. His protagonist Sherlock Holmes was the very epitome of cold-blooded analysis and deduction, no hunches or spirit rapping for him! Rationality rules Holmes' thought processes: "From a drop of water, a logician could infer the possibility of an Atlantic or a Niagara without having seen or heard of one or the other." Yet Doyle, his creator, was about as gullible an occultist as anyone, before or since, going so far as to enthusiastically endorse the infamous "Cottingley fairy"

Dante and Beatrice gaze upon the highest Heaven, The Empyrean, per Dante's *The Divine Comedy.* Illustration by Gustave Doré, 1832–1883. PUBLIC DOMAIN

photographs, taken by two girls in Yorkshire, England, who snapped themselves with pictures of fairies they'd cut out from books.

Likewise, the late nineteenth-century philosopher William James, father of "radical empiricism," a hardheaded rationalist by day and credulous fool by night, was convinced of the reality of the supernatural realm by a smooth-talking medium, Leonora Piper. Mediums back then encouraged the departed to rap on tables, toot trumpets, and write reassuring messages on slate boards. ("I'm fine. Food's great. Send money.")

Or Thomas Edison, personification of the pragmatic inventor-businessman—and loyal patron of Madame Blavatsky, founder of the esoteric Theosophical Society.

Going back even further, if anyone can be called the father of the Scientific Revolution, it's Isaac Newton: physicist, mathematician—and alchemist and believer in the apocalypse. Historians believe that Newton's scientific work was less important to him than the supernatural. The economist-philosopher John Maynard Keynes summed up his life: "Newton was not the first of the age of reason, he was the last of the magicians."

I could go on and on, but the question remains: lacking any hard evidence, why do so many of us—69 percent of Americans, according to a recent Pew poll—"absolutely believe in heaven"? Are our brains naturally wired to believe in eternal life? Or is it simply that we're having such a good time here that the idea of shuffling off the mortal coil becomes too great a burden to bear?

For me, I can see spending a year or two on the Other Side, but *forever*? Seems like an awfully long time to fill. Those harps, you know.

78

ANTARCTIC ATLANTIS

Looking for Plato's mythical land

Never mind Waldo or Bernie. *Where's Atlantis?* More trees have probably been felled in the service of books and magazines speculating about the site of the "lost civilization" than about any other pseudoscientific boondoggle. The mythical land has been located all over. One popular candidate is the small Greek island of Santorini, aka Thera, remnant of a massive volcanic blast dated to about 1600 BC known as the "Minoan eruption." Although the location doesn't fit—Atlantis is supposed to have been *west* of the Mediterranean, not within it—the fact that an explosion four times the intensity of Krakatoa virtually annihilated an entire island resonates with Plato's original story. Other favored locations include Troy, the Dogger Bank in the North Sea (which was land until rising sea levels flooded it around 6000 BC), the Canary Islands, Greenland, and (of course) the so-called Bermuda Triangle. Plus, you've got to give props to another imaginative idea, that the ruins of Atlantis live on, under the thick ice cap that covers the continent of Antarctica.

To the best of our knowledge, the ancient Greek philosopher Plato, writing around 360 BC, started the whole shenanigans. In one of his dialogues, he tells us that beyond the Pillars of Hercules (where the Mediterranean enters the Atlantic Ocean between Gibraltar and Morocco), "there lay an island which was larger than Libya and Asia together. . . .

Now in this island of Atlantis there existed a confederation of kings, of great and marvelous power, which held sway over all the island."

In the next dialogue, Plato has his great-uncle Critias recount the fall of Atlantis 9,000 years earlier, after the Atlanteans had conquered most of the Mediterranean basin. Having defeated Libya, Egypt, and much of southern Europe, they attacked Athens. Bad move. Not only did the Athenians fight them off, but the vengeful Athenian gods sent "portentous earthquakes and floods, and one grievous day and night befell them, when the whole body of your warriors was swallowed up by the earth, and the island of Atlantis in like manner was swallowed up by the sea and vanished."

Most historians believe that Plato's Atlantis was a complete fabrication, especially because the philosopher dates its fall to some 7,000 years before the emergence of Athens as a city-state. An exception is the late scholar and former U.S. Navy captain Arlington H. Mallery, who proposed that a map found in 1929 in Istanbul's Topkapi Museum (the former Imperial Palace) shows an *ice-free* Antarctica south of South America.

The map was drawn by Turkish adventurer Piri (= admiral) Reis in 1513, just two decades after Columbus "discovered" the New World. It incorporates information from Columbus, Vasco de Gama, and other early European navigators. Most other researchers believe that what Mallery took to be the Antarctic coast is simply the result of Reis trying to fit his information to the shape of the parchment. I agree; I think we can safely assume he was making up that part of his map. For example, "Antarctica" is shown connected to South America, complete with a picture of a snake and the note "large snakes have been found, for which reason the Portuguese infidels [that is, Christians] have not landed in these lands which are said to be very hot." The real Antarctica wasn't discovered until 1820.

However wacky, Mallery's speculations were all it took for Italian engineer Flavio Barbiero to embrace the theory. According to him, Reis must have had access to earlier maps drawn when Antarctica was ice-free, which he claimed in his book *A Civilization under Ice* was only 10,000 years ago. (It's actually 34 million years.) The "civilization" in the title is, of course, Atlantis. In Barbiero's imaginative version of history, a few

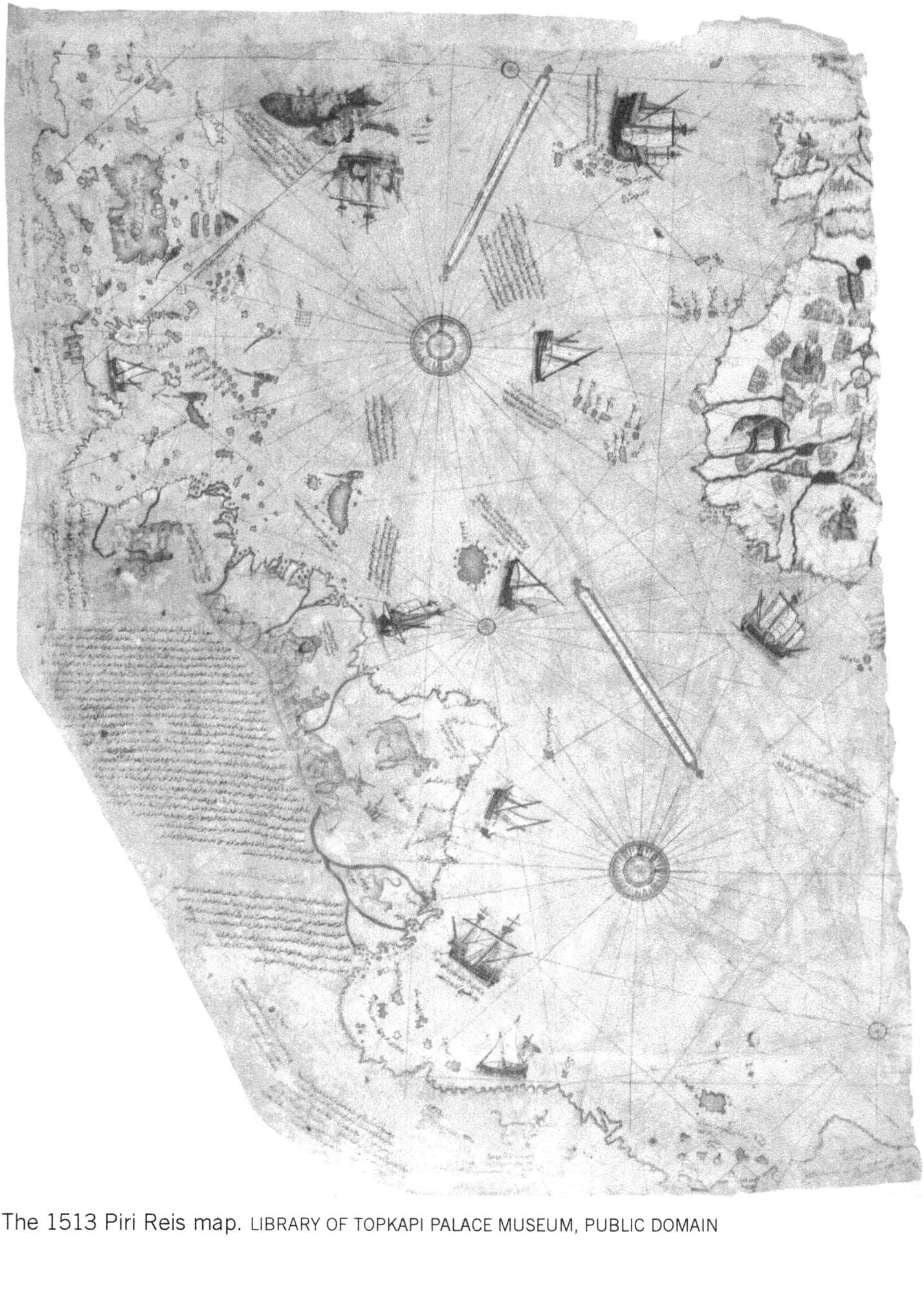

The 1513 Piri Reis map. LIBRARY OF TOPKAPI PALACE MUSEUM, PUBLIC DOMAIN millennia ago, Earth's axis was tilted differently, so today's polar regions enjoyed mild climates. Unfortunately for the inhabitants of those parts, including Atlanteans, a comet or meteor knocked the axis off-kilter to where it is today, and Atlantis was buried by the subsequent snowfall. So all we've got to do now, according to Barbiero and several copycat writers following his lead, is to dig down through the mile-plus thickness of ice, and you'll find Atlantis.

In another, grim retelling of Plato, the Nazis appropriated Atlantis for their own ideology of racial superiority, building on the writing of Madame Helena Petrovna Blavatsky, cofounder of the Theosophical Society (precursor of the New Age movement). Dictated in Atlantis (where else?), her 1888 book *The Secret Doctrine* illuminated the evolution of entire races, including the genesis of superior Aryans. Nazi propagandists ran with her hogwash, using it to justify the genocide of "inferior" races.

Like the Garden of Eden, Atlantis lives on in legend, each generation adding another layer to the story: it's going to be around, more or less, forever.

79

THE MYTH OF THE GOOD OLD DAYS

They weren't

When the future is dark, people turn to an idealized past, a lost golden age. Nostalgia becomes a refuge against dangers, a cocoon against the announced declines.

—Jean-Louis Missika, former deputy mayor of Paris

Writing in about 700 BC, Greek poet Hesiod waxed nostalgic for the "good old days" as he imagined them: "They lived as if they were gods, their hearts free from all sorrow. . . . The fruitful grainland yielded its harvest to them of its own accord." In reality, his—and our—good old days sucked. Whether we're talking in terms of decades, centuries, or millennia, any comparison between life *now* and life *then* overwhelmingly favors the former; we're far, far better off now than ever. Consider the following:

- The United States was built on the twin evils of genocide of indigenous people and slavery of Africans. Lynching was a fact of southern life less than 100 years ago. ("Progressive" Theodore Roosevelt could get away with, *"I don't go so far as to think the only good Indians are dead Indians, but I believe nine out of ten are."*)
- In 1980, the U.S. murder rate was 10.2 per 100,000 people annually; now it's about half that (and about one-thirtieth what it was in Europe in the Middle Ages).

- In 1900, infectious diseases caused half of all deaths in the United States. Today, it's less than 3 percent. The average lifespan of a baby born in the United States in 1900 was 49; now it's 78. In the period 1990–2010, the percentage of the global population living in what the World Health Organization defines as "extreme poverty" decreased from 43 to 21 percent and continues to fall.
- War deaths have massively declined. The first half of the twentieth century saw *tens of millions* of world war deaths; by the second half, that was down to *millions* in superpower "proxy wars"; in this century, our worst wars, such as those in Iraq, Syria, and Ukraine, kill *hundreds of thousands*. Nuclear weapons haven't been used since 1945. Despite the brutally high, racially biased incarceration rates in this country, we've come a long way from old-time punishment. Here's how Assyrian king Ashurnasirpal (884–859 BC) dealt with rebels: "I flayed all the chiefs who had revolted. . . . Some I impaled upon the pillar on stakes. . . . Some I cut off their noses, ears and fingers; of many I put out their eyes."
- Nobel savages' (Rousseau's phrase) stewardship of the land is sometimes compared favorably to our ravaging of the environment, but those so-called gentle people hunted mammoths, dodos, and moas to extinction; turned the Valley of Mexico into a dust bowl, and chopped down all the trees in Iceland and on Easter Island.

Where did our predilection to romanticize the past and condemn the present come from? Two (of many) reasons: our 24/7 "if it bleeds it leads" (ledes?) news keeps death and destruction in our awareness, and we're genetically programmed to pay more attention to negative than to positive data (it was better to assume that rustling in the tall grass was a saber-toothed tiger even if it turned out to be wind).

Of course, we've got overwhelming challenges ahead: climate change, overpopulation, environmental devastation, resource scarcity, ocean acidification, a billion people still living in poverty, not to mention Gaza, Ukraine, and Eritrea, but still, talk of the "good old days" is a canard. On average, humans have never been healthier or more prosperous, long-lived, peaceful, or democratic.

80

KING ARTHUR

Once and future king

Arthur long sustained his sinking country, and roused the broken courage of its citizens for war.

—*Deeds of the Kings of the English*,
William of Malmesbury, ca. 1125

I used to give one-evening community college classes on such eclectic topics as "The Seven Wonders," "Dinosaurs," "Mars," and "King Arthur." Attendance was unpredictable—except for "King Arthur," which *always* attracted a full house. What I was seeing wasn't new. Since the Middle Ages, the Arthurian legends have consistently captured large audiences with their tales of chivalry and romance. Here, I'll try to explain their timeless popularity.

Around 1200, medieval French poet Jean Bodel classified the storytelling themes of his age into three "Matters." The oldest was the "Matter of Rome," which included legendary Greek tales of the Trojan War and the subsequent founding of Rome by Aeneas, together with exploits of real-life military strongmen, such as Alexander and Julius Caesar. The youngest, the "Matter of France," concerned the French king Charlemagne and his 12 knightly "peers," including Roland (as in the classic *Chanson de Roland*). And the third was the "Matter of Britain": the story of King Arthur.

You might think that a figure as well known in popular culture as Arthur, together with his court at Camelot, the Round Table of knights, and the quest for the Holy Grail, must have at least *some* basis in fact. Yet historians debate whether a real-life Arthur actually existed or, if he did, which of maybe six questionable contenders was the prototype. That's because it all happened in post-Roman Britain, during the era often, somewhat mistakenly, referred to as the "Dark Ages," roughly from AD 400 to 700, when European historical records are particularly sparse (see "The Myth of the Dark Ages").

DUX BELLORUM

Assuming that he was a real person, Arthur lived in late fifth-century Britain, but not until three centuries later is he mentioned by name as a *dux bellorum* (warlord). Romanized Britain had enjoyed a roughly three-century era of peace and prosperity in the period before Arthur,

Tintagel Castle, birthplace of King Arthur according to Geoffrey of Monmouth. The dates don't fit since the castle here dates back only to medieval times. AUTHOR

from Emperor Claudius's conquest in 144 to the evacuation of the last Roman troops in 410. They left behind a British populace "entirely ignorant of the whole practice of war," quoting the only British historian of that time, the monk Gildas. With the Romans gone, outsiders stepped up the raids: Picts and Scots from northern Britain and Saxons, Angles, and Jutes from present-day Denmark and northern Germany. According to Gildas, the land was devastated with fire "until it burnt nearly the whole surface of the island, and licked the Western Ocean with its red and savage tongue." What the country needed was a leader capable of rallying the Britons to defend the land from the invaders—maybe someone named Arthur?

The Welsh monk Nennius, writing his *History of the Britons* in about AD 830, certainly thought so: "Then Arthur along with the kings of Britain fought against [the Saxons] in those days, but Arthur himself was the warlord. . . . The twelfth battle was on Mount Badon in which there fell in one day 960 men from one charge by Arthur; and no one struck them down except Arthur himself, and in all the wars he emerged as victor."

Archaeologists are certain that the entire country didn't immediately sink into anarchy after the legions left, pointing to such sites as the Roman cities of Dorchester and Wroxeter, where civilized town life continued for at least another 200 years. Also, many Iron Age (ca. 600 BC–AD 100) forts were rebuilt in the late fifth century, evidence of organized resistance. The earthworks of Somerset's South Cadbury, for example, were strengthened around AD 470, and digs there have unearthed Mediterranean pottery and remains of Roman-style buildings from that time. More telling, perhaps, is the curious pattern of continental-style graves and pottery, abundant in the lowlands of southern and eastern Britain beyond London from before 500 and again after 600. But there's a century-long gap, perhaps indicating a period when Anglo-Saxon invaders were held at bay?

MOUNT BADON

Which brings us to the Battle of Mount Badon, "the last great victory of the fatherland," according to the monk Gildas, writing around 540. Tellingly, perhaps, he doesn't mention Arthur. (That was left to Nennius.)

The "Great Work" at Viroconium, present-day Wroxeter. Originally a Roman crossroads city, it was abandoned when the legions left Britain, only to be resurrected during the so-called Dark Ages. AUTHOR

Historians are pretty sure that at least one major battle was fought around AD 500 in which Britons decisively beat the invaders, but they are divided on where it happened and whether a British (or Romano-British) general called Arthur led the defending forces.

One prominent theory holds that "Mount Badon" was Badbury Rings, a large earthen Iron Age fort located in Dorset, in southwestern England. Badbury was one of the strongholds refortified around AD 500, and that, together with the similarity of the names "Badon-Badbury," has led some historians to propose that a warrior named Arthur led British forces to a decisive victory there, in what Alfred, Lord Tennyson poetically called "the last dim weird battle of the west."

Another source of Arthurian lore is the *Annals of Wales*, written around 1100 but referring to events long before. Two entries in particular are of interest: one alludes to the "Battle of Badon," while the other tersely notes, "The fight at Camlann in which Arthur and Medraut were killed." Medraut (or Mordred) is usually painted as Arthur's enemy, but here it's ambiguous. One theory is that Camlann is Camboglanna (meaning "crooked glen"), now known as Birdsowald, a Roman fort on Hadrian's Wall in northern England. Below the fort, the river Irthing makes a sharp bend. Was Arthur not a king or *dux bellorum* but merely one of two local tribal leaders battling it out in post-Roman Britain?

GEOFFREY AND CHRETIEN

Following Nennius's account of a superhuman warrior killing 960 enemies, Arthur became fair game for storytellers. By 1135, Geoffrey of Monmouth had elevated him to king of Britain (he was only a *dux bellorum*, or warlord, in Nennius's account), son of Uther Pendragon, legendary Welsh king of post-Roman Britain. After Geoffrey's Arthur marries Guinevere, daughter of a noble Romano-British family, he established his court at Caerleon in Wales, once a major Roman fortress (and incidentally where Tennyson wrote his own version of the tale, *Idylls of the King*).

But it was French writer Chretien de Troyes who inspired the full flowering of the Arthur legend. Writing around 1180, Chretien pulled together previous strands of the story, including tales that probably came to him via the Celtic-British colony of Brittany ("little Britain") on mainland Europe. Knowing a good plotline (anticipating soap operas by many centuries), he brought us the "eternal triangle" theme of Arthur/Guinevere/Lancelot, in which Guinevere betrays her husband the king by consummating her love for his favorite knight, Lancelot. Chretien also introduces the element of Christianity in the form of the Holy Grail and mentions, for the first time, Arthur's castle: Camelot.

MALORY

While great medieval romances abounded, such as those of Alexander the Great and King Charlemagne, none were milked to the extent we see with King Arthur in the "Matter of Britain." The most imaginative of the repackagers was English writer Thomas Malory (1405–1471), who combined the French courtly-love angle with Arthurian Welsh sources, creating in *Le Morte d'Arthur* (The Death of Arthur) a story about believ-

Merlin shows Arthur the ruins of Camelot in Gustave Doré's 1868 illustration for Tennyson's *Idylls of the King*.
PUBLIC DOMAIN

able and fallible heroes. Malory's account was one of the first books to be printed in England, by the pioneer printer William Caxton in 1485. It's here that Arthur is given his delicious sobriquet: "many men say there ys wrytten uppon [Arthur's] tumbe thys: *Hic iacet Arthurus, rex quondam rex futurus*": "Here lies Arthur, the once and future king." (This was the title adopted by T. H. White in his four-part romance novel about Arthur.)

Malory's interpretation, bolstered by Tennyson's jingoistic Victorian-era retelling *Morte d'Arthur*, not to mention the musical *Camelot* with its associations with John F. Kennedy, is the one that we know today and that our grandchildren and their children will know tomorrow. Arthur's tale is as timeless as the art of storytelling.

81

THE GREAT FLOOD

The reality behind the Noah story

> *Tear down thy house, build a ship; abandon wealth, seek after life; scorn possessions, save thy life. Bring up the seed of all kinds of living things into the ship which thou shalt build.*
>
> —COMMAND FROM THE GODS TO UTNAPISHTIM, *EPIC OF GILGAMESH*

Similarities between the flood legend in the Babylonian *Gilgamesh* saga (parts of which date to 4,000 years ago) and the much later story of Noah's ark (in the Old Testament Book of Genesis) are striking. Both Noah and Utnapishtim are divinely commanded to save themselves and their families, together with representatives of all living animals, by building a huge boat in which to survive the coming deluge. Once the boats (or arks) are ready, torrential rains flood Earth entirely, killing all living creatures. When the waters finally recede, Noah's ark lands on Turkey's Mount Ararat, while Utnapishtim's boat comes to rest on Mount Nisir, probably Pir Omar Gudrun in Kurdistan.

The truth is that there isn't nearly enough water on Earth to fulfill the claim, made by both Gilgamesh and Genesis, of a *worldwide* flood. Even if all the frozen water presently locked up in our planet's ice sheets and glaciers melted, sea level would rise by only (!) 260 feet. Still, it's a fair question to ask whether a real-life catastrophic flood could

The Deluge. Frontispiece to Gustave Doré's illustrated edition of the Bible, 1866.
PUBLIC DOMAIN

have prompted both these two ancient stories and other similar deluge accounts from the Middle East.

The most convincing argument for a historical antecedent I've seen came in 1997, when geologist Bill Ryan and geophysicist Walter Pitman, both at Columbia University, proposed that a massive deluge occurred in the Black Sea region nearly eight millennia ago. Before that time, the Black Sea was a freshwater lake. When rivers feeding it changed course, evaporation resulted in the water level falling to some 300 feet lower than the nearby Mediterranean Sea. In their scenario, the Mediterranean breached a sill of soft rock, opening up the Bosphorus channel between the two seas and flooding a vast area of land—60,000 square miles—in a few months. The researchers postulate that about

10 cubic miles of water "poured through each day, two hundred times what flows over Niagara Falls," wiping out any unfortunate farmers and herders then living in the low-lying regions.

Since that time, geologists from around the world have taken up the gauntlet thrown down by Ryan and Pitman. Roughly speaking, two rival camps have emerged. While one agrees with their "catastrophe" version, the other claims that any enlargement of the Black Sea occurred much more sedately, with minimal loss of life.

What I find so intriguing is the possibility that the writers of *Gilgamesh* and Genesis documented an actual event so terrifying that an unbroken line of storytellers kept its memory alive, in exaggerated form, for thousands of years. If the debate turns out favorably for Ryan and Pitman, that should tell us something about the staying power of oral tradition.

82

A BRIEF HISTORY OF HELL

The myth to end all myths

In order that the happiness of the saints [in heaven] may be more delightful to them and that they may render more copious thanks to God for it, they are allowed to see perfectly the sufferings of the damned [in hell].

—Thomas Aquinas, *Summa Theologica*

The invisibility of the ground beneath our feet has led to many wild speculations about its nature, including ancient myths of a subterranean realm where you went when you died. The oldest of these tales has *everyone* ending up down there, while later beliefs limit inhabitants of the underground world to those who had led wicked lives. Today, most Americans are in no doubt: a 2023 Gallup survey reports that 58 percent of us believe in hell, while heaven scores 69 percent.

Starting with what is arguably the oldest surviving work of literature, the 5,000-year-old Babylonian *Epic of Gilgamesh*, it's pretty obvious that those who dwell in the underworld are not happy campers. For example, Gilgamesh's pal Enkidu has a terrifying dream portending his own death and subsequent journey down to "the house of darkness . . . the house where one who goes in never comes out again; the road that, if one takes it, one never comes back; the house that, if one lives there, one never sees light; the place where they live on dust, their food is mud." No internet either.

Charon herding sinners onto his boat en route to hell. Gustave Doré's illustration to Dante's *Inferno*, 1861. PUBLIC DOMAIN

Hades, predecessor of the Christian hell, was originally the name given to the god who ruled those "infernal regions." That god later morphed into Pluto, a tougher character, one suspects, than Disney's canine goofball. Hades was a real place for the ancients, including the Roman writer Virgil, who assured his readers that its entrance was near Mount Vesuvius and where "the whole country is cleft with chasms, from which sulfurous flames arise, while the ground is shaken with pent-up vapors, and mysterious sounds issue from the bowels of the earth."

Hades was a place of perpetual torment and frustration. Crafty Sisyphus, for example, was forced to eternally push a huge boulder up a hill (it kept rolling back down) as punishment for trying to outsmart Zeus. Meanwhile, Tantalus (whence our word *tantalize*) was damned to eternal thirst and hunger. Every time he bent down to drink the water in which he stood, it receded. When he reached for fruit above him, it pulled away.

Where Hades was cold and dark, hell was hot and bright with flaming brimstone, that is, sulfur. As portrayed in Christianity, Islam, Judaism, and Zoroastrianism, it's the final dwelling place of the dead after they've been condemned at the Last Judgment. The traditional Christian notion of hell seems to stem from the Jewish myth in which ancient Jerusalem's city dump, Ge (valley) Hinnom, is transformed into the fiery Gehenna. From there, it was but a short step to the blazing domain of Satan and his evil angels, enthusiastically portrayed by such artists as Hieronymus Bosch, William Blake, and Gustav Doré, in which it looks like one hell of a place.

83

THE LURE OF LABYRINTHS

From Minotaur's den to a meditative walk

Standing at the center of the world's best-known labyrinth in Chartres cathedral a few years ago, I regretted that I was a couple of centuries too late to check out the brass plate connecting Chartres with the Greek myth of the labyrinth. Only the metal studs now remain from where an engraving used to show the struggle between Theseus and the half man/half bull Minotaur, the plate having been melted down for cannons during the French Revolution. The surrounding labyrinth, fashioned from contrasting colors of polished granite, is an intricate 11-circuit path that takes about 20 minutes to walk slowly.

According to legend, the original labyrinth was built by the craftsman Daedalus, who had been commissioned by King Minos of Crete to create an escape-proof prison for the king's stepson, aka the Minotaur. Long story short: the god Poseidon gave Minos a white bull, and instead of sacrificing it to the god, the king kept it for himself. This infuriated Poseidon, who—with a little help from love goddess Aphrodite—made Minos's wife lust for the bull. She then gives birth to the Minotaur.

In Ovid's telling (in *Metamorphoses*), Daedalus constructed such a complicated labyrinth that he was barely able to escape it himself. Later, the hero Theseus entered the labyrinth, unwinding a ball of wool given him by Minos's daughter Ariadne. Happy Hollywood ending: he kills the Minotaur and follows the thread to find his way out and back to Ariadne.

The 861-foot-long, 11-circuit labyrinth in Chartres cathedral was completed around 1220. AUTHOR

So originally, a labyrinth (literally *place of the double axe*, referring to the "logo" of the palace of Knossos in Crete) was a complicated maze. Today, however, a labyrinth is more likely to mean a meandering route for meditation or prayer: you just follow the single path, whereas a maze is a confusing network of interconnecting pathways designed to get you lost.

You don't have to go to France to experience the "Chartres design" since replicas abound. If you're in San Francisco, for example, check out the one in Grace Cathedral, created in 1994. According to the cathedral's website, "The labyrinth is an archetype, a divine imprint, found in all religious traditions in various forms around the world. By walking a replica of the Chartres labyrinth . . . we are rediscovering a long-forgotten mystical tradition."

As an aside, Chartres cathedral was barely saved from destruction in 1944 thanks to the initiative and bravery of an American army officer. The Allies were preparing to shell the cathedral, believing that it was being used as a German observation post, when Colonel Welborn Griffith Jr. volunteered to go behind enemy lines with a single enlisted man to check. He reported back that the cathedral was clear of German troops, and the order was withdrawn. Griffith died in action soon after.

84

ATTILA, SCOURGE OF GOD—AND DIPLOMAT

Beyond the myth

When in 2020 wealthy businesswoman Kelly Loeffler advertised herself as "more conservative than Attila the Hun" in a runoff election for a Georgia Senate seat (she held the seat for just three weeks—it's complicated), she was doing the leader of the Huns a disservice. Unless she was planning on wholesale slaughter, she could hardly make a dent in the reputation of the so-called Scourge of God. Loeffler wasn't the first to use the "right of Attila" line. It's been claimed by or used against, among others, linguist William Safire and conservative politicians Barry Goldwater, Ronald Reagan, and Margaret Thatcher. In Attila's time, "left" and "right" in politics didn't mean much. Back then, politics essentially boiled down to, "Were you an invader or a defender?" Attila was an invader, no doubt, as opposed to the Romans, who were occupied mostly with defending their crumbling empire.

So who was Attila? We actually know very little about him, and what we do know is pretty tainted since nearly all the sources we have for his life were written by his enemies, of whom there were many. History, it's said, is written by the victors, but in this case, it was written by those who could write, the Huns themselves probably being illiterate.

What we know is that Attila was born early in the fifth century AD—probably 406—and died about 47 years later, either from a hemorrhage while celebrating his latest marriage or perhaps by assassination.

Extent of the Hunnish Empire in AD 450. ADAPTED FROM USER ANDREI NICU BY SLOVENSKI VOLK VIA WIKIMEDIA, CREATIVE COMMONS

Attila and his elder brother Bleda had succeeded their uncle, Rugila, as joint leaders of the Huns. After the early death of Bleda, Attila ruled alone, for a total of nearly 20 years as coleader or sole leader. During his reign, he unified several other powerful tribes in eastern and central Europe, including the Ostrogoths, the Bulgars, and the Alans, bringing them under his rule. In this, he seems to have been as much a diplomat as a warlord. (Perhaps he inspired Otto von Bismarck to do the same more than 1,000 years later, when Bismarck unified the Germanic independent states under the leadership of Prussia?)

Attila lived during the last days of the Western Roman Empire, a time when, increasingly, foreign mercenaries were employed by the state to protect the frontiers and deals were made to pay what was essentially protection money to tribes that threatened the *Pax Romana*. At the time, Rome and its provinces were constantly under threat from Germanic tribes to the north and from nomadic tribes to the east, whose homeland was central Asia and the Caucasus. The Huns seem to have been the most powerful of these to the extent that Attila was able to threaten first the Eastern Roman Empire, centered in Constantinople, and later the Western Empire based in Rome. By all accounts, they were expert bowmen on horseback, giving them an edge over Roman infantry in battle. One Roman historian wrote that they "are almost glued to their horses," while another said they "live and sleep on their horses."

Led by Attila and after many battles and skirmishes, the Huns were finally defeated—or at least brought to a standstill—at the Battle of the Catalaunian Plains in northeastern France on June 20, 451, when the Huns faced a combined force of Romans and Visigoths. Within two years, Attila was dead, and soon after, the Huns disappeared from history. Almost. In the Middle Ages, several Hungarian writers portrayed the Huns positively as their "glorious ancestors."

So, Scourge of God or brilliant statesman? Either way, labeling him as an arch-conservative minimizes what we do know about him. Too bad almost all our information comes from those who had the most to gain by vilifying him; I'd have been happy to buy him a beer and listen to his side of the story.

85

THE MYTH OF THE DARK AGES

Time to drop the label

All the objects shone in the sunshine as on the day they were buried.
—Basil Brown's diary, August 1939

Britain's "Dark Ages" may have formally ended in a field in Suffolk, England, on July 21, 1939. That's when archaeologist Peggy Piggott uncovered a tiny gold pyramid encrusted with garnets from the Sutton Hoo ship burial. Within days, excavators went on to discover a fabulous Anglo-Saxon hoard of precious metal objects that had been buried alongside a local king, Raedwald, within a 90-foot-long ship. He died in AD 624, thus dating the site to 1,500 years ago.

The so-called Dark Ages weren't dark anymore! In one stroke, Sutton Hoo changed archaeologists' understanding of early English culture. If the inhabitants of fifth-century Britain (in the middle of the "darkest of the Dark Ages," according to the *Oxford Illustrated Histories*) were capable of making or acquiring by trade such gorgeously intricate objects buried with their king, they can hardly be considered as anything other than enlightened. Check out *The Dig*, a dramatized version of the find with Ralph Fiennes as Basil Brown, the working-class "excavator," and Carrie Mulligan (heart, be still) as landowner Edith Pretty. And if and when you visit London, treat yourself to a day at the British Museum, allowing a full hour upstairs in Room 41, which houses the European

Two of nearly 300 items from the Sutton Hoo hoard now on display in the British Museum. Top: Gold garnet shoulder clasp encrusted with Sri Lankan garnets. ROBROY/CREATIVE COMMONS/WIKIMEDIA. Below: Gold belt buckle, weighing nearly one pound. JONONMAC46/CREATIVE COMMONS/WIKIMEDIA

Medieval collection. The centerpiece—literally—of this room is the Sutton Hoo ship burial, with nearly 300 pieces on display.

To be fair, almost no historians now use the term *Dark Ages*, which originated with the worldly Tuscan scholar Petrarch, writing in the 1330s. Believing himself to be living in an age of darkness (despite the Renaissance!), he looked back longingly to the enlightened and cultured "classic ages" of ancient Greece and Rome. Later writers identified Dark Age Europe as the time between about AD 500 and 1100, when records were few and classical learning all but lost. The fall of Rome in AD 476 is usually given as the start of the collapse, even though subsequent Germanic kings maintained many of the Roman traditions (including the Senate), while the Eastern Roman Empire, centered in Constantinople, survived for another millennium.

Modern historians, in avoiding the term *Dark Ages*, refer instead to the Early Middle Ages. Not only do we now know much more about those times from archaeology, paleography, and other fields of study, but ask yourself, does an age that gave us Charlemagne, the Abbasid "Golden Age of Islam," Beowulf, the *Book of Kells*, Peter Abelard, Hildegard of Bingen, and Avicenna—not to mention the Sutton Hoo hoard—really deserve the label "dark"? And for that matter, were the Greeks and Romans (dysfunctional, slaveholding, misogynistic, warring) all that enlightened?

NUMBERS

The magic of math

86

THE NUMBER OF THE BEAST

666 or 616?

Quick, what's the sum of the numbers on a 36-slot roulette wheel? You could add 1 to 2 to 3, all the way up to 36. Or you could note that 1 + 36 = 2 + 35 = . . . = 18 + 19. That is, 37, 18 times over, or 666. And 666, as everyone knows, is the Number of the Beast—or was until 1999, when 666 lost some of its notoriety and became, well, just another number.

A quick diversion here: 666 is actually a pretty cool "triangular" number. Imagine filling a triangle with balls, as if you're racking a game of pool. The first row holds one ball; the second holds two, for a total of 3; the third holds three, total 6; and the fourth holds four, total 10, giving the first four triangular numbers: 1, 3, 6, 10. The thirty-sixth triangular number (36 again!) is 666.

The "beastly" connection of 666 comes from a riddle in chapter 13, verse 18, of the last, crazy, book of the Bible, Revelation. The King James Version of the Bible reads, "Here is wisdom. Let him that hath understanding count the number of the beast: for it is the number of a man; and his number is Six hundred threescore and six." This chapter previously described a vision of two horned beasts, the first of which—the one referred to in verse 18—has seven heads with blasphemous (from the Christian point of view) names, including standard titles given to Roman emperors: "Lord and Savior" and "Savior of the World."

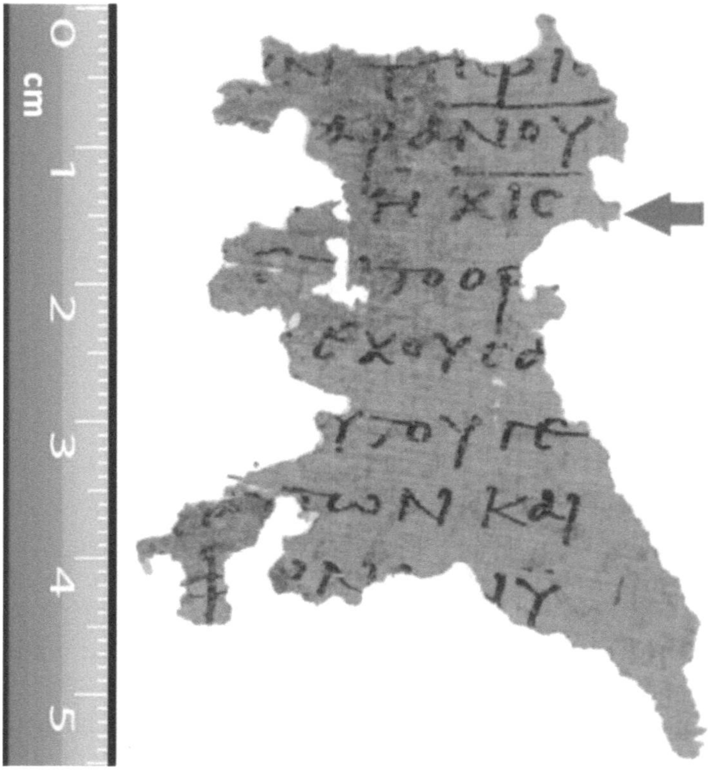

Papyrus fragment "P. Oxy. 4499," now at Oxford University, is the oldest witness to Revelation 13:18. The third line gives 616 for the Number of the Beast. WIKIMEDIA COMMONS

Although the King James Version gives the Number of the Beast as 666, as do most modern translations (not to mention *The DaVinci Code*, *The Omen*, and countless other references in popular culture), 616 has long been an also-ran. The *Novum Testamentum Graece* (Greek New Testament) of 1514, for example, has 616, not 666. More importantly, 616 is found in the very oldest fragment of Revelation, dating to about AD 300, found at the archaeologically rich site of Oxyrhynchus in southern Egypt. Unearthed in 1895, the fragment was illegible until the advent of ultraviolet imaging. When published in 1999, it pretty much put the kibosh on the 666 tradition. The number is in the third line: chi-iota-sigma, or hexakosiai-deka-hex: 616.

So, 666 or 616? It doesn't really matter since both numbers probably refer to one particular "beast," the Roman emperor Nero, who reigned from AD 54 to 68. John, author of the Book of Revelation (sometimes called the Apocalypse, or "unveiling"), recorded his visions while living in a cave on the Aegean island of Patmos, possibly during or soon after the reign of Nero, archenemy of the new Christian faith. Having visited his cave and looked out on nothing but bucolic green fields, I have to wonder if he'd ingested some of the local magic mushrooms.

According to Ellen Aitken, professor of early Christian history at McGill, Revelation is actually a thinly disguised political tract, "a critique of the politics and societies of the Roman Empire . . . written in coded language and riddles." Transliterating Nero's Greek name and title "Neron Caesar" into Hebrew (in which letters stand for numbers) gives 666. Doing the same with the Latin version, "Nro Caesar," gives 616.

So that tasteful "666" tat you got on your butt during your goth phase? Sorry, 616 is the new 666. And the roulette wheel no longer symbolizes Nero the Beast, who, according to most historians, wasn't that bad a guy after all.

87

METERING THE METER

First, go to the North Pole

We're spatial creatures. We live in a three-dimensional world, with our fourth "dimension"—time—conceptualized in terms of space. That is, we refer to time as if it *were* space: we turn clocks "forward" or "back" twice a year, we agree to "move up" a meeting, and we wonder "where" the time "went." In short, we live embodied lives in three dimensions of "real" space and one of "pseudo" space. This essay is about how we measure that space. It all comes down to the meter.

Here in the United States (along with Liberia and Myanmar), we pretend we haven't adopted that newfangled metric system (tell that to STEM folks). More to the point, our standard units of length—inch, feet, yard, and mile—are actually defined in metric terms. The inch, for example, is *exactly* 2.54 centimeters, a centimeter being one-hundredth of a meter, which is an improvement over when an inch was legally "three grains of barley, dry and round, placed end to end, lengthwise."

How the meter itself came about is another story, with its origins in the French Revolution. Along with the short-lived 10-hour day (100 minutes per hour, 100 seconds per minute), the utopian visionaries of the National Convention in Paris declared in 1791 that the metre/meter (Greek μετρέω/*metreo*, "to measure") was defined as "one ten-millionth of the distance from the equator to the North Pole." Previous attempts to tie the meter to the length of a pendulum with a half period of one

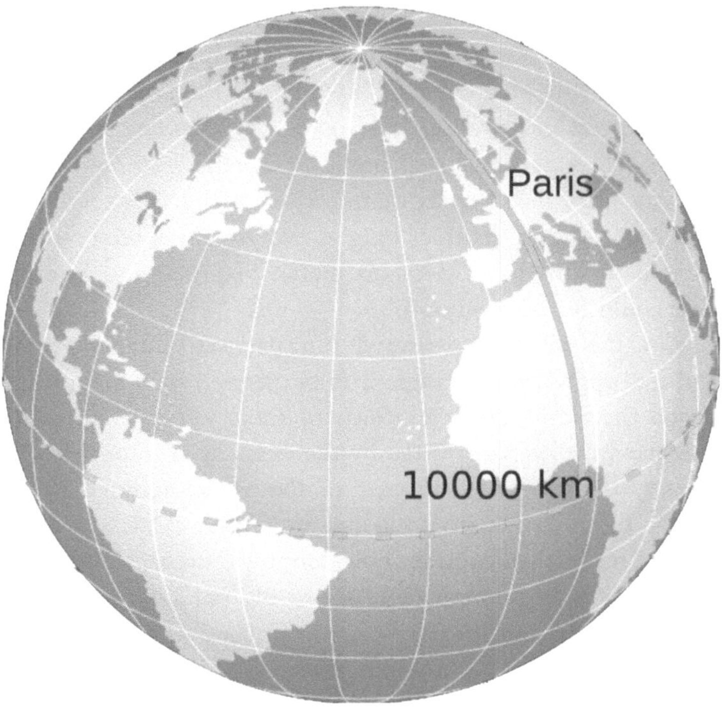

In 1791, the meter was defined as one ten-millionth of the distance from the equator to the North Pole. WIKIMEDIA COMMONS

second had faltered when it was realized that the period changes from place to place due to variations in local gravity.

Fortunately, you don't have to literally measure the distance from the North Pole to the equator to use this new definition. You can extrapolate, measuring the difference in latitude (obtained from the height of the sun at solar noon) between two known points on the same meridian of longitude. In 1791, the French Academy of Sciences commissioned astronomers Jean Baptiste Joseph Delambre and Pierre Méchain to measure the exact distance between Dunkirk to Barcelona in order to determine the length of the meridian arc that runs roughly through Paris.

Two years later, based on the astronomers' results, France—quickly followed by most other European countries—adopted the standard meter in the form of an etched iron bar kept in a Paris vault. No matter that Méchain was later found to have fudged his results and that the prototype meter bar was actually 0.02 percent short because of a miscalculation of Earth's oblateness ("flattening" of the poles). You can read about it in Ken Alder's riveting 2002 book *The Measure of All Things: The Seven-Year Odyssey and Hidden Error That Transformed the World.* As it happened, the error didn't really matter. With man being the measure of all things (according to man), as long as everyone was now on the same system based on this new "standard meter," national and international commerce could function smoothly.

In 1876, the newly constituted International Bureau of Weights and Measures upgraded the single iron bar to 30 platinum-iridium bars kept around the world. Then, in 1960, the meter was redefined in terms of a particular wavelength emitted by the element krypton when heated in a vacuum. The modern definition of the meter, adopted in 1983, is the distance light travels in a vacuum in 1/299,792,458 of a second. Sharp-witted readers will wonder how defining a universal unit of length in terms of time is an improvement on the old pendulum measurement. Fortunately, modern cesium clocks come to the rescue since they routinely measure time to 1 part in 10^{13} power.

It will probably take another revolution to adopt the metric system in the United States, so meanwhile, we're stuck with inches, feet, yards, and miles, which, I guess, is why God gave us Siri and Alexa to help us out.

88

MEASURING OUR LIVES

60 minutes, 24 hours, 7 days, 52 weeks

Hodgepodge. Mishmash. Dog's breakfast. How else to describe the many and various systems we blithely use to measure everything from time to gasoline to dress sizes to Starbucks coffee? We divide our pounds into sixteenths, pizzas into eighths, dollars into hundredths, hours into sixtieths, weeks into sevenths, whiskey into fifths, days into twenty-fourths (but clocks—and feet—into twelfths), football games into quarters, baseball games into ninths, circles into three hundred-sixtieths, reams of paper into five-hundredths (20 quires of 25 sheets each), and on and on. Be thankful you don't live in the United Kingdom, where your weight comes in stones and pounds, 14 pounds to a stone, and where there are not 100 but 112 pounds in a hundredweight!

Dividing things into halves—the binary system—is probably the oldest such mechanism. For example (though I'm not sure anyone actually uses all of these anymore), 1 tun = 2 pipes = 4 hogsheads = 8 barrels = 16 kilderkins = 32 bushels = 64 demi-bushels = 128 pecks = 256 gallons = 512 pottles = 1,024 quarts = 2,048 pints = 4,096 chopins = 8,192 gills. Notice the doubling each time. To further confuse matters, "British Imperial" measures are about 20 percent larger than their U.S. equivalents. That is, except for the imperial fluid ounce, which is about 4 percent smaller. Go figure. To help you in your calculations, you can start with this useful tidbit: the American gallon is based on "Queen Anne's Gallon" of 231 cubic inches.

The Babylonians famously gave us the base-60 system, source of our seconds and minutes (both temporal and angular), which makes all sorts of sense when you're looking to divvy up, say, sheaves of grain (without scales). You can easily divide 60 sheaves into halves, thirds, quarters, fifths, sixths, tenths, twentieths, and thirtieths without incurring a remainder. The decimal system, complete with a zero, came later thanks to Indian mathematicians—handy for beings with 10 digits (or 20 if you count your toes, the vigesimal base adopted by the Mayans) but not so useful for dividing your pile into quarters and thirds, which is probably why we're still stuck with 12 inches in a foot and 24 hours in a day, with eight hours of sleep, one-third of our existence, supposedly being the norm.

Whatever we may think of it, our current mess isn't going to go away soon. Look what happened when, in the middle of their revolution in the late 1700s, the National Assembly of France tried to rationalize time by dividing each day into 10 hours of 100 minutes each. Adopted in October 1793, the system lasted barely a year before the tried-and-true 60 minutes/24 hours was readopted. The Revolutionary Calendar—three weeks of 10 days each in a month, with a year lasting 12 months plus a bit—fared slightly better until Napoleon dumped it 12 years later.

The takeaway from all this is our willingness to just go with the flow, barely noticing what a kludge it all is. Like irregular verbs and laws that make no sense, we accept these systems and pass them on to our children no matter how messy, irrational, or confusing they are.

89

ARCHIMEDES' TOMB

Anticipating calculus by 2,000 years

And so I took a good look round all the numerous tombs that stand beside the Agrigentine Gate. Finally I noted a little column just visible, surmounted by a sphere and a cylinder.

—Cicero

The sphere and the cylinder were the tip-off for the Roman writer Cicero as he searched for Archimedes' tomb 137 years after the Greek mathematician's death. These objects were carved on his tomb to celebrate what reputedly was Archimedes' proudest achievement: calculating the volume of a sphere. It's relatively easy if you know how to integrate, but Archimedes lived 2,000 years before Isaac Newton and Gottfried Leibniz independently (probably) invented calculus. How did he do it? He used calculus, masquerading as "infinitesimals."

Quick bio: Archimedes of Syracuse (287–212 BC) wasn't only the greatest mathematician of the ancient world but also a physicist, engineer, inventor, and astronomer. He died by accident, sort of. After the Romans finally prevailed following a two-year siege of Syracuse, Sicily, a Roman infantryman found Archimedes working on a problem in the sand outside his house. "Don't disturb my figure!" growled the old man, prompting the impatient soldier to slay him.

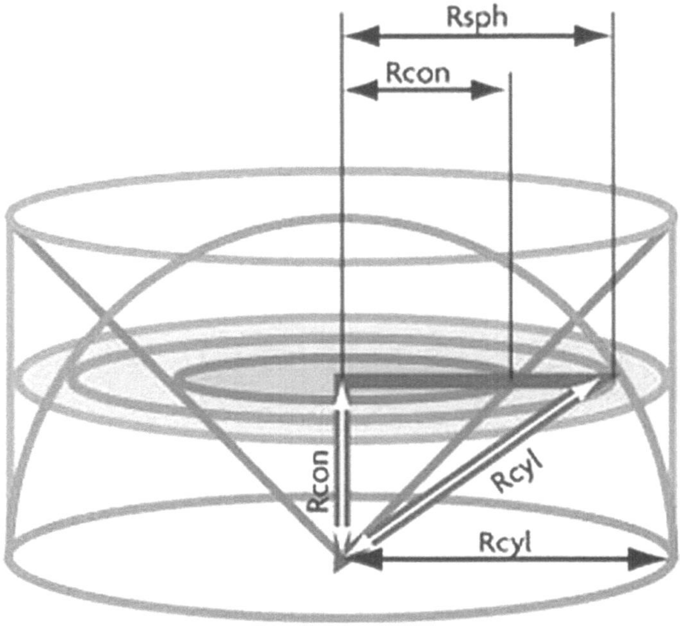

Volumes of the cylinder, hemisphere, and cone are in a 3:2:1 ratio.
AUTHOR/*NORTH COAST JOURNAL*

His proof for the volume of a sphere is nothing short of brilliant and well worth a couple of minutes of your time to follow. Notice that a cylinder that perfectly encloses a hemisphere of radius r will have both radius and height r, as will a cone fitted inside (here upside down). Any plane parallel to the ends of the cylinder, like the one shown, cuts through the cylinder, hemisphere, and cone. Let the radius of the three circles so formed be *Rcyl*, *Rsph*, and *Rcon*, respectively:

- In the right-angle triangle (double lines), Pythagoras' theorem tells us that $Rcyl^2 = Rsph^2 + Rcon^2$.
- Since a circle's area is proportional to the square of its radius, the area of the cylinder's circle = area of the hemisphere's circle + area of the cone's circle.

So far, so pretty straightforward. Archimedes then took his "anticipating calculus" leap of faith. Reasoning that the plane was equivalent to a thin (*infinitely* thin!) slice, he extrapolated from plane to volume:

- Volume of the cylinder's slice = volume of the hemisphere's slice + volume of the cone's slice.

Summing an infinite number of infinitely thin slices (remember "integration" in your high school calculus class?), he concluded the following:

- Volume of cylinder ($\pi r^2 h$) = volume of hemisphere + volume of cone ($1/3\ \pi r^2 h$), thus giving the volume of a sphere as $4/3\ \pi r^3$.
- From which, as I bet Archimedes was thrilled to discover, the ratio of the volumes of cylinder:hemisphere:cone is 3:2:1.

Tomb-topping magic!

90

ROMAN AND ARABIC NUMERALS

Our debt to Leonardo of Pisa

C-bill, World War I, Pope Leo XIV, Super Bowl XLIII, Final Fantasy II. Although we normally use so-called Arabic numerals, their Roman counterparts are far from dead. In addition to the above examples, they're used to date movies; to identify book volumes, such as encyclopedias; to mark off the front matter in books from the body, usually in lowercase (i, ii, iii, iv); for the cornerstones of buildings; and for subsections in laws (Title XV). Once you start noticing them, you see them everywhere.

The Romans got their number system from the Etruscans of ancient Italy, Etruscan 1, 5, and 10 being written as I, Λ, and X, hence Roman I, V, and X. The Etrusco-Roman symbols probably derived from tally sticks. A notch on a stick became I, while every fifth notch was marked with a double cut (Λ or V), and the tenth notch was crosscut (X). (Or perhaps 10—the number of digits on two hands—came first and the Λ or V was subsequently derived from half an X.)

The other Roman numerals (L, C, D, and M, for 50, 100, 500, and 1,000) don't lend themselves quite so readily to interpretation, but we're pretty sure that the Romans ultimately adopted C for *centum* (100) and M for *mille* (1,000), while the D may have come from *demi-mille* (half a thousand). The Roman system doesn't include a zero, which is probably why the Arabic system (which originated in India) was adopted so readily in the Middle Ages.

Top: Two styles of numerals, same street number at Talisman in Eureka, California. AUTHOR. Bottom: Public phone keypad, Cairo. COURTESY PETE HAMILTON

One problem, of many, with Roman numerals is that the system isn't totally additive. An additive system would, for example, express the year 1944 as MDCCCCXXXXIIII, but instead it's usually shortened to MCMXLIV. The symbols C, X, and I each precede a symbol of higher value, so the result is interpreted as 1,000 (M) plus 1,000 - 100 (CM) plus 50 - 10 (XL) plus 5 - 1 (IV).

So why do old clocks with Roman numerals show 4 as IIII, not IV? Nine is shown as IX, so shouldn't 4 be written as IV? One explanation is that the pairs of numerals IV and VI and IX and XI are easily confused but that IV and VI are doubly confusing because the IV is partly upside down. Clockmakers responded to this ambiguity by using IIII for 4.

While you're pondering that, you can ask yourself, why do clocks only go up to 12 when there are 24 hours in a day?

The hero who introduced the Arabic system to the West (including the Indian "discovery" of zero) was the greatest mathematician of the Middle Ages, Leonardo of Pisa, aka Fibonacci (1170–1250). Arabic numbers from zero to nine are ٩ ٨ ٧ ٦ ٥ ٤ ٣ ٢ ١ ٠, on which Fibonacci based his symbols 0 1 2 3 4 5 6 7 8 9. He did so much more, though. His book *Liber Abaci* (Book of Calculation) was seminal in converting Europe from the clumsy Roman numbers, with their complicated arithmetic, to the lean, clean, mean system we use today. Compared to the Roman CCXIV, our "positional" decimal system of ٢١٤, or 214 (two 100s plus one 10 plus four 1s) looks elegantly simple—once you know how it's done.

Thanks Fibonacci!

91

HALLOWEEN, SAMHAIN, JACK-O'-LANTERN

Our pagan festivals

The Christian celebration of Halloween (holy evening), like the pagan Celtic festival of Samhain, is that liminal time of year when the boundary between the worlds of the living and the dead is thin. For our ancestors, that's when ghosts and spirits were propitiated with offerings of crops, milk, and wine. Today, these ancient customs from Ireland, Scotland, and Wales (bastions of Celtic heritage) are mirrored by such modern counterparts as *Dia de los Muertos* (Day of the Dead), when Latinos leave food and drink at the graves of departed family members.

In pre-Christian Britain and Ireland, Samhain (pronounced SAH-win), literally "summer's end," marked the beginning of the winter season and, according to some historians, the start of the new year. Crops were brought in, cattle brought down from the hills, livestock slaughtered (as summer grass died), and bonfires lit, and precursors of modern trick-or-treaters went from house to house in costume demanding gifts in return for reciting poetry. Sound familiar?

The Feast of All Hallows, or All Saints, was an attempt by the early church to lessen the importance of pagan festivals by substituting them with Christian holy days, like pegging Christmas to the Roman solstice festival of Saturnalia. In the case of Samhain, in 835, Pope Gregory IV designated November 1, the traditional date of Samhain, as All Saints' Day.

And just as the Yule log reminds us of the pagan roots of Christmas, our carved candlelit pumpkin, the jack-o'-lantern, predates an earlier tradition. The story goes that Jack was a ne'er-do-well hard drinker whose misdeeds precluded happiness in the afterlife. Staggering home after a boozy night, Jack was ambushed by Old Nick himself. "Follow me to purgatory," he ordered. Jack complained that he would need a stick "from yonder sycamore" if he were to walk all the way to hell. Grudgingly, Satan climbed the tree to break off a branch, at which Jack trapped him by quickly carving a cross in the bark. So a deal was struck: Jack would free him if Satan promised to not claim his soul. When Jack died, Satan kept his word but instead tossed him a blazing lump of coal straight from the fires of hell. It being a cold night, Jack put the coal into a scooped-out turnip for warmth and light. And today, Jack and his lantern still roam between heaven and hell as he seeks a place to rest.

Somehow, this old Irish story traveled to East Anglia, in England, where jack-o'-lantern became another name for will-o'-the-wisp, the local name for *ignis fatuus*—foolish fire—the odd phenomenon of light appearing at night above marshy ground. It's not well understood but is probably caused by the spontaneous ignition of methane and phosphene from rotting vegetation. *Ignis fatuus* is rarely seen today, with the disappearance of both wetlands and the dark of night. Have you seen it, perchance?

Immigrants from the Old World brought Halloween over to the United States, where it caught on like wildfire despite objections from the Puritans. Next Halloween, as you light your pumpkin candle or give those pesky neighbor kids a handful of tooth-destroying candies, know that you're maintaining a custom whose origins come from long ago and far away.

92

A FUNNY THING HAPPENED TO PI

Finding its ten-trillionth digit

In what follows, I claim that the history of pi, or π (approximately 3.14159), the ratio of a circle's circumference to its diameter, parallels, in a rough way, the entire history of mathematics.

Four thousand years ago, the only way to get a value for π was to physically measure it: how many times does a diameter length of rope fit around a circle's circumference? Three and a bit? Close enough back then. Around 2000 BC, the ancient Egyptians were using a value of 256/81 (= 3.1604 . . .), which they derived by assuming that a circle with a diameter of 9 units has the same area as an 8 × 8 square, which was certainly better than exactly 3, per the Old Testament (I Kings 7:23).

Around 250 BC, Archimedes of Syracuse (*Eureka!*) first figured out how to calculate π using the geometry of regular polygons, noting that the circumference of a circle lies between the perimeter lengths of its inscribed and circumscribed polygons. With enormous effort, working with a 96-sided polygon, he showed that π lies between $3\frac{10}{71}$ and $3\frac{1}{7}$ (which averages to approximately 3.14159). Archimedes' method was still being used in Europe in the 1600s, with very little improvement in accuracy.

The next big step for mathematics—and for the calculation of π—probably originated in the south of India around AD 1400 with the discovery of *infinite series* by Madhava. While infinity had scared off earlier mathematicians, Madhava and (nearly three centuries later) Scotsman James Gregory and German Gottfried von Leibniz independently

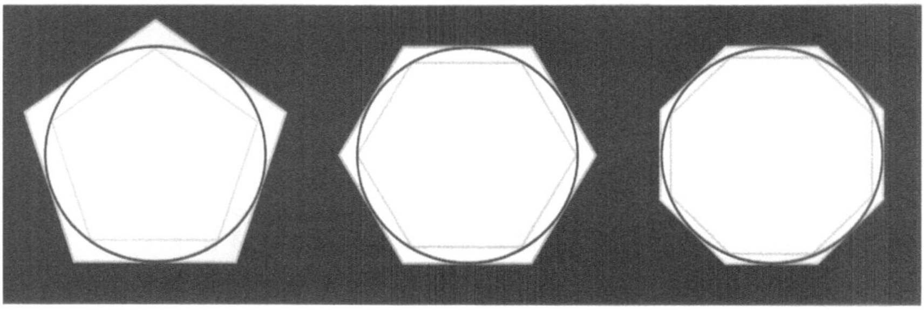

Archimedes' method for calculating π. The length of the circumference of a circle lies between the perimeters of inscribed and circumscribed polygons: the more sides, the better the accuracy. Working with a 96-sided polygon, Archimedes computed π to an accuracy of 1 part in 10,000. LESZEK KRUPINSKI, WIKIMEDIA COMMONS

showed how to calculate π to any desired degree of accuracy by repeated application of the simple, infinite formula $\pi/4 = 1 - 1/3 + 1/5 - 1/7 + 1/9 \ldots$ Sounds great, except it's insanely inefficient. Calculating this to half a million terms yields only five decimal places of accuracy!

In 1706, English astronomer John Machin improved on the Madhava-Gregory-Leibniz discovery: same idea but much better formula. I'm particularly partial to Machin's formula because it's incorporated into the 99-cent "PiAcademic" app. With it, I can compute (I use the word loosely) the first 65,000 decimal digits of π in 18 seconds on my iPad.

Just 30 years ago and still paralleling the general evolution of mathematics, Machin-type formulae for pi gave way to "iterative algorithms," in which the same function is applied to successive iterations; that is, each subsequent iteration yields a closer approximation to pi than the previous value. Common to all these schemes, from Archimedes on, was the requirement that to know, say, the millionth digit of π, you needed to know all previous 999,999 digits. Makes sense, right?

Not to Québécois mathematician Simon Plouffe, who figured out a "spigot formula" for pi in 1995, so called because it calculates a single isolated digit, similar to focusing on a single drip from a faucet rather than the entire flow. If you'd told previous pi calculators that it was possible to figure out the ten-trillionth digit of pi (in the 1s and 0s of binary code) knowing none of the previous digits, they'd have laughed at you.

But that's just what Plouffe showed the world of mathematics. It's 0, by the way.

SPACE

Earth's-eye view of the universe

93

EINSTEIN, NEWTON, AND THE ECLIPSE OF 1919

Fudging the results

What follows is the tale of the high-stakes test of a scientific theory that made headlines around the world more than 100 years ago in the aftermath of World War I, linking the lives of two pacifists on opposing sides of the conflict: German-born theoretician Albert Einstein and English astronomer Arthur Eddington. Einstein, already famous for his groundbreaking 1905 special theory of relativity, published the general theory of relativity in 1916.

In a nutshell, mass causes the "fabric" of space to be warped, which in turn bends light. Instead of Isaac Newton's spooky conception of gravity as "invisible forces at a distance," Einstein's gravity was simply our perception of local mass-caused warping ("curvature") of the fabric of space. Now space, far from being empty, was imbued everywhere with this property of curvature, measurable in the same way one could measure the salinity of a drop of ocean water at any place and depth.

Such a quick summary hardly does justice to the complexity of the general theory, with its three-dimensional geometry, tensor mechanisms, and field equations. The story goes that when a colleague suggested to Eddington, director of the Cambridge Observatory in the United Kingdom, that probably only three people in the world understood the theory, he paused, then replied, "I'm trying to think who the third person might be."

The players: Albert Einstein, Arthur Eddington, and Frank Dyson. EINSTEIN: F. SCHMUTZER/ANZENBER GALLERY; EDDINGTON: GEORGE GRANTHAM BAIN/LIBRARY OF CONGRESS; DYSON: BAIN NEWS SERVICE/LIBRARY OF CONGRESS, ALL PUBLIC DOMAIN

Following the publication of Einstein's 1916 paper, British Astronomer Royal Frank Dyson realized that the solar eclipse of May 29, 1919, would present an opportunity to test Einstein's revolutionary "light-bending" predictions by measuring the deflection of starlight passing very close to the sun. Normally, of course, we can't see stars close to the sun, that is, in broad daylight—except during a total solar eclipse, when the moon obscures the sun's disk. The 1919 eclipse was particularly propitious because the sun would be "in" the bright Hyades star cluster at the time of the eclipse, making for a particularly good photo opportunity.

At the time, few scientists doubted that the sun would bend starlight. Einstein himself had said so as early as 1911, based on his special theory. Light, in this view, has mass, and mass is attracted to mass per Newton. The mass of the sun would cause the light from stars close to the sun to be slightly bent. How slightly? About 0.86 seconds of arc, equivalent to measuring the width of a penny from more than two miles away.

Four years later, Einstein—having figured out general relativity (with its curvature of space)—predicted a new value, 1.75 seconds of arc. All that was needed to decide between Newton's 200-year-old theory of gravity and Einstein's new theory was to snap a few photos during the eclipse and check whether the sun deflected starlight by Newton's 0.86 or Einstein's 1.75 seconds of arc. This proved to be a far more difficult project than anyone could have anticipated.

Note the irony here: seven months after Britain and Germany ended their bloody four-year conflict, British astronomers were to confirm that German-born Einstein's theory of gravity would supersede that of Englishman Isaac Newton.

Britain launched two expeditions to observe the total eclipse of May 29, 1919. One was from the Greenwich Observatory, under the direction of Frank Dyson: "He was at that time very skeptical about [Einstein's] theory, though deeply interested in it; and he realized its very great importance," wrote Arthur Eddington, director of the Cambridge Observatory, which sent the second expedition. In contrast to the skeptical Dyson, Eddington was Einstein's great champion in Britain. He saw the upcoming eclipse not only as a way to vindicate his trust in the man and his theory but also (as a pacifist) as an opportunity to promote postwar rapprochement of scientists from opposing sides of the conflict. Coming right after the bloodiest war ever fought, this idealistic notion met with resistance in many quarters. It's to Dyson's and Eddington's credit that the expeditions happened at all.

THE EXPEDITIONS

The armistice of November 1918 ("the eleventh hour of the eleventh day of the eleventh month") couldn't have come early enough for two scientists. While Dyson appointed deputies to take the crucial photographs from Sobral in northern Brazil, Eddington himself led the Cambridge group to the Atlantic island of Principe, off the coast of Africa. Anyone who has ever viewed or tried to view a total eclipse of the sun will understand why *two* expeditions were sent out: the weather can make or break the occasion. In the event, both parties obtained photographs (actually, glass plates) of relatively bright stars whose light grazed the sun during the eclipse. Photographs: yes. Good photos, sharp photos, unambiguous photos: no. While the cloud-free Sobral site yielded 27 photos of varying quality from two telescopes, Principe's one telescope delivered just two usable photos out of a total of 16 due to cloud cover.

The job was far from over. In order to "weigh light" and settle the Newton-versus-Einstein debate, experts in Greenwich and Cambridge compared the eclipse results with photos of the same star fields at night, that is, when their light wasn't deflected by the sun's mass. This turned out to be enormously complicated due to (1) the microscopic differences in star positions between the eclipse and reference plates; (2) overexposure of many of the Sobral plates, resulting in stars appearing as blobs rather than sharp pinpoints; (3) focusing problems with one of the Sobral instruments due to the temperature drop during the eclipse; and (4) the difficulty of obtaining valid reference plates.

Cutting to the chase, *The Times* of London headline for November 7, 1919, read, "Revolution in Science: New Theory of the Universe: Newtonian Ideas Overthrown," which neatly sums up the take-home lesson from the eclipse expeditions. Einstein was vindicated, the measured deflection agreeing with his prediction of 1.75 seconds of arc. Newton's gravity was passé.

THE CONTROVERSY

What wasn't publicly discussed at the time is that data from one of the three sets of observations (two from Sobral, one from Principe) were eliminated from consideration and that this excised dataset *supported the Newtonian explanation of gravity*. At the time, few quibbled with the publicized results, and for 60 years, that's more or less how things stood. Officially, Eddington's and Dyson's expeditions had confirmed Einstein's theory of general relativity.

Until other ways of testing general relativity had been devised, almost no astronomers voiced what, in retrospect, is obvious: measuring the deflection of starlight near the sun during a total eclipse is a *terrible* way of testing general relativity! It's inexact; it depends on perfect seeing conditions; it can be thrown off by moisture in the atmosphere, refraction, and 100 other variables; and it's subject to interpretation by fallible (and probably biased) humans. Only in the 1970s did radio telescopes easily verify Einstein's prediction for space curvature by measuring the deflection of radio signals (detectable in broad daylight) from distant quasars.

FUDGED RESULTS?

In 1980, historians John Earman and Clark Glymour published a scholarly paper in *Historical Studies in the Physical Sciences* in which they took

Eclipse instruments at Sobral, 1919. C. DAVIDSON, PUBLIC DOMAIN

a long, hard look at the events preceding the November 1919 announcement of the results of the expeditions. They concluded that Eddington, in his zeal to champion what he had called Einstein's "beautiful and profound theory," had fudged the results, tossing out the "bad" (that is, counter to Einstein) dataset because it didn't fit his preconceived beliefs. Had these data been used, the conclusion from the 1919 eclipse expeditions would have been much more nuanced, with no clear winner in the Newton-versus-Einstein "duel in the sun." Subsequently, the popular science book *Einstein's Luck* by John Waller popularized this revisionist history. Waller's title sums up the new thesis: Einstein was lucky. The publicized results from the eclipse could easily have gone the other way had not Eddington been biased in favor of Einstein.

Fortunately for the history of science and Eddington's reputation, not everyone was wholly satisfied with Earman and Glymour's version of the 1919 events. Physicist Daniel Kennefick from the University of Arkansas undertook a re-reinterpretation and concluded that Eddington was not involved in the decision to discount the pro-Newton, anti-Einstein data (https://arxiv.org/abs/0709.0685).

The Sobral expedition had been launched by the Greenwich Observatory, and it was there, under the supervision of the skeptical Frank Dyson *independently of Eddington*, that the decision was made. The telescope in question had lost focus during the eclipse (probably due to a sharp drop in temperature), resulting in what was, in 1919, an insurmountable scaling problem.

THE IRONY

One final irony: Kennefick discovered that the problem plates had been reanalyzed, without fanfare, in 1979 using modern computerized astrometric (star-measuring) equipment. Remember that Einstein's theory predicted a 1.75-second displacement for starlight "grazing" the sun, about twice that of the old Newtonian theory of gravity. Dyson's team at Greenwich had measured a mean shift from the out-of-focus—and subsequently discarded—data of 0.93 seconds, favoring Newton. When reanalyzed with modern equipment (which eliminated most sources of error), the shift was now an Einstein-compatible 1.55 plus or minus 0.34 seconds!

In other words, if those doughty astronomers of 1919 had magically been given access to today's tools, they would have seen that all three sets of data unambiguously supported Einstein's "spacetime" model of gravity.

94

THE MONK'S TALE

When the moon split in two

Four centuries after being burned at the stake on an otherwise chilly February morning in 1600, the freethinking Dominican monk Giordano Bruno is still creating controversy. Actually, it's the formation of the lunar crater named for him that's causing all the debate.

According to the English monk Fratello Gervase of Canterbury (1141–1210), an extraordinary event took place on the early evening of June 25, 1178 (our calendar), the likes of which has never been seen since: "A marvelous phenomenon witnessed by some five or more men who were sitting there facing the [two-day-old thin crescent] Moon. . . . Suddenly the upper horn split in two. From the midpoint of this division, a flaming torch sprang up, spewing out . . . fire, hot coals and sparks . . . the body of the Moon which was below writhed, as if it were in anxiety, and to put it in the words of those who reported it to me and saw it with their own eyes, the Moon throbbed like a wounded snake."

What did they see? According to a proposal made in 1976 by geologist Jack Hartung, it was the impact from a *huge* (two miles across, perhaps) meteor or asteroid, resulting in the formation of the Giordano Bruno lunar crater. Immediately after impact, molten dust would have obscured part of the crescent moon; that is, the "horn" would have been "split." From Gervase's description, the impact occurred close to the right edge of the moon, halfway between the moon's equator and its North Pole. Crater Bruno, on the far side just out of sight from us, at latitude

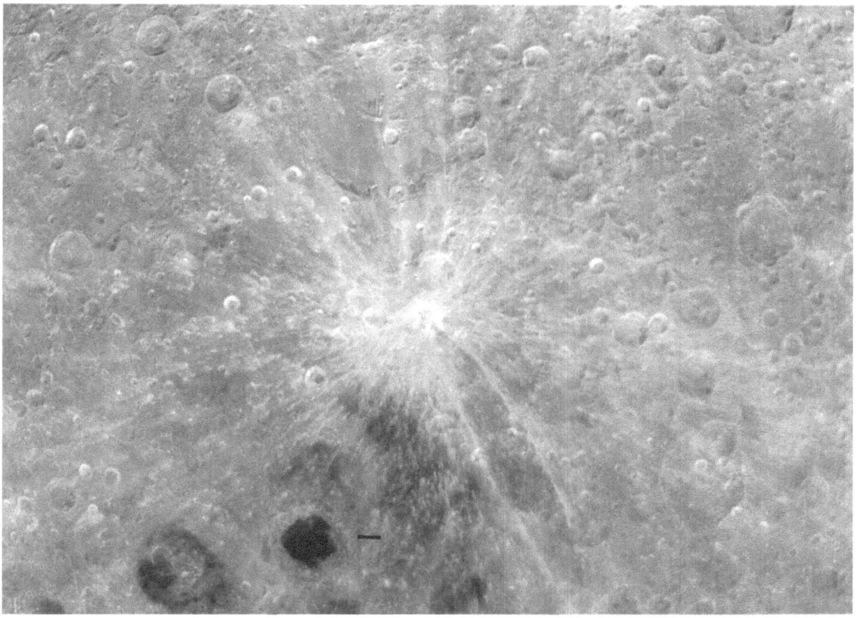

Crater Giordano Bruno lies at the center of pristine rays of ejecta up to 100 miles long. NASA

36 degrees north, fits nicely. What's particularly compelling about Hartung's proposal is Bruno's extreme youth, probably being the youngest crater of its size on the moon. We know this because the rays spreading out from the crater, formed of light-colored ejecta material, are pristine, while those from older craters have been degraded by the constant rain of micrometeorites on the lunar surface. And at 14 miles in diameter, Bruno is certainly consistent with the eruption of a vast cloud of ejecta visible from Earth.

And that's the problem, according to lunar expert Paul Withers of the University of Arizona. An impact of that magnitude would have catapulted some 10 million tons of debris into Earth's atmosphere, causing, at the very least, a weeklong meteor storm, spectacular enough to be recorded by someone, somewhere. Yet none of the astronomical records from the time—European, Chinese, Arabic, Japanese, and Korean—mention what Withers claims would have been the "best fireworks show in history."

So there's the conundrum. If the Bruno crater was caused by a huge impact in 1178, why didn't anyone record the subsequent earthly meteor shower? The answer will have to wait for future astronauts to ship samples from the crater back for analysis so that its precise age can be calculated. If those samples show that Bruno was formed about 800 years ago, we can be pretty confident that Gervase recorded the birth of a lunar crater, thus solving the mystery of a moon that "writhed" and "throbbed" on that long-ago midsummer night.

95

WERE THE OLD DAYS SHORTER?

Or am I getting older?

Yes, and yes. Really, days are getting longer. The energy to run our tides has to come from somewhere—no magic perpetual motion machine—and that somewhere is Earth's angular momentum, or spin. The moon generates tides gravitationally by pulling most strongly on the side of Earth facing it, causing the planet (especially the malleable oceans) to bulge. Since our planet rotates faster, once every 24 hours, than the moon's 29.5-day "synodic" orbit, the tidal bulge travels ahead of the moon, tugging the moon forward. The net result is that Earth imparts energy to the moon, speeding the moon up in its orbit (so it moves away from us) and slowing down Earth's spin. The change isn't a lot, but it adds up. Each century, tides generated by the moon and, to a lesser extent, the sun add about 1.7 milliseconds to the length of a day, while the moon moves about 13 feet away from us.

At least three pieces of evidence confirm this:

- Coral clocks: Corals go through daily and seasonal growing cycles, creating bands (like growth rings in trees) that record the number of days in a year. Around 350 million years ago, according to this "clock," a year was 385 days long, and a day lasted about 23 hours. Additionally, in southern Australia, tidal "rhythmites" (specks of minerals deposited daily in sandstone) imply a 400-day year about 620 million years ago.

Earth and moon from Mars. The moon, with 1/81 of the mass of Earth, slows Earth's rotation, making the length of the day increase about 2 milliseconds per century. HiRISE camera on the Mars Reconnaissance Orbiter, 2007. NASA/JPL

- Lasers beamed at the moon: In the late 1960s and early 1970s, U.S. astronauts and unmanned Russian probes placed laser reflectors on the moon's surface. Changes in the round-trip time of a laser beam confirm the 13-foot annual recession.
- Ancient eclipse records: Early records of solar eclipses confirm that days are getting longer. An article by Richard Stephenson in the October 1982 *Scientific American* lists 28 historical solar eclipse records, mostly Chinese, including one dating back to 709 BC. By averaging them out, Stephenson was able to show that days have been getting longer by about 1.78 ± 0.11 milliseconds per century over the past three millennia. (Stephenson was just the man for the job, being both an emeritus professor in the Department of Physics department and the Department of East Asian Studies at Durham University in the United Kingdom.)

Extrapolating backwards, immediately after the moon was formed about 4.5 billion years ago, it was probably just four Earth radii away from us (compared with 15 times that now), and Earth's day was six hours long. Looking ahead, the moon will keep receding as Earth's spin continues to slow down. Eventually, they'll become tidally locked, meaning that only one hemisphere of our planet will see the moon. (The moon is already tidally locked, so only one hemisphere of the moon sees Earth.) At that time, a day on Earth might last 1,000 hours or more.

Nothing to get too excited about just yet. It won't happen for another 50 billion years.

96

IMPACT! 50,000 BC

Arizona's Barringer Crater

"Whatever created this hole was one scary mother," I thought while standing on the rim of what's billed as the "first proven, best-preserved, meteor crater on earth." It's so obviously formed by a meteorite (a meteorite is a meteor that makes it to Earth before being vaporized)—perfectly round, steep sided—that it might have been imported directly from the moon. Yet until 50 years ago, most scientists believed it was the caldera of an ancient volcano.

We now know that some 50,000 years ago, a lump of nickel-iron, weighing one-third of a million tons, slammed into what was then a grassy plain 40 miles east of present-day Flagstaff, Arizona. The explosion would have dazzled—right before vaporizing—any woolly mammoths or North American camels in the area. Equivalent to some 150 Hiroshima-size atomic bombs, the impact scattered hundreds of millions of tons of rock onto the surrounding countryside. About half the bulk of the meteorite was vaporized, leaving—in theory—a rich lode of metal waiting to be discovered and mined.

That was what mining engineer Daniel Barringer believed, anyway, when President Theodore Roosevelt signed a patent granting him the rights to a square mile of land around the center of the crater in 1903. For the next 25 years, Barringer's Standard Iron Company searched fruitlessly for the remaining metal, hoping to extract it and make huge profits for the shareholders. Years after those efforts were abandoned, the Barringer

Barringer Crater from the southeast. Fortunately, the meteorite just missed the visitor center on the right. USGS

family made the best of it by turning the huge crater into a tourist attraction, and today, you can tour the excellent visitor center and take a guided walk along part of the north rim.

Very little of the actual meteorite has ever been found, which is why the volcanic origin was once favored. Not until 1960, when planetary scientist Eugene Shoemaker identified rare forms of dense silica (created only by nuclear explosions or impacts) at the bottom of the crater, was the meteorite origin completely accepted.

My wife and I spent a fine couple of hours in the museum and along the crater rim, trying (not very successfully) to imagine the force of the explosion that rent such a vast crater in our planet's crust not so long ago. It's well worth a visit: five stars.

The three-quarter-ton Holsinger Meteorite in the visitor center is the largest fragment found. Authentic human shown for scale. AUTHOR

97

GALILEO'S MOONS

Everything changed one night in 1610

Galileo, perhaps more than any other single person, was responsible for the birth of modern science.

—Stephen Hawking

On the night of January 7, 1610, our understanding of the cosmos changed forever. That's when an obscure professor from Italy's University of Padua, Galileo Galilei, using his homemade 20-power telescope for the first time, saw three points of light near the planet Jupiter. He wrote in his journal about "three fixed stars, totally invisible [to the naked eye] by their smallness, close to Jupiter and lying on a straight line through it."

Night by night for a week, he saw that they were moving, and he realized that they weren't fixed stars but, rather, moons orbiting the giant planet. At the time, the idea that any celestial body (other than Earth) could have satellites flew in the face of the Aristotelian view of the heavens. Aristotle's Earth-centered universe still ruled in early seventeenth-century Europe despite the publication of Copernicus' sun-centered theory in 1543.

Galileo had discovered three of what we now call the four "Galilean moons" of Jupiter: Io, Europa, and Callisto. He found the fourth, Ganymede, the following week, guessing correctly that it had been eclipsed by Jupiter when he first spotted the other moons.

Jupiter and the four Galilean moons, November 27, 2024. Left to right Callisto, Europa, Io, and Ganymede. COURTESY GRACE WHEELER

In addition to the cosmic, not to mention religious, implications supporting the Copernican model, Galileo recognized the practical importance of Jupiter's moons in giving us a "clock in the sky" that could resolve the problem of longitude (see "The Longitude Challenge"). Longitude—essentially the difference between local time and Greenwich time—could now be measured from anywhere on Earth by knowing the predicted eclipse times of Jupiter's moons (which you can find, for example, in *Astronomy* magazine).

Suppose you know from the tables that Io will eclipse at exactly 6:00 a.m. Greenwich time, that is, longitude zero) and your clock is set to local time, established by setting it to noon, when the sun is at its highest point. And say you observe the eclipse at 9:44 p.m., that is, a difference of 8 hours 16 minutes from Greenwich time. That's equivalent to 124 degrees west, which happens to be the longitude of my hometown: Eureka, California.

98

DINOSAURS DIED, MAMMALS THRIVED

An ancient fluke put us here

Ten minutes before a Manhattan Island–size space rock barreled through Earth's atmosphere 66 million years ago, the ecological balance between dinosaurs and mammals was working just fine. Until the fiery end of the Mesozoic Era, the two groups had been living and thriving alongside each other for tens of millions of years, each mostly keeping to their own ecological niches. Mammals stayed small, no larger than today's raccoons, while dinosaurs grew to enormous proportions. A titanosaur, for example, could weigh up to 100 tons. Mammals essentially filled the "small is beautiful" niche, into which, in their absence, some terrestrial dinosaurs might have evolved. (Avian dinosaurs did, of course: birds.)

This state of affairs—big dinosaurs, small mammals—lasted throughout most of the Mesozoic Era. Dinosaurs appeared about 225 million years ago, while the common ancestor of today's mammals dates to around 160 million years ago. The arrangement seems to have been beneficial to both groups, although there were exceptions. For example, baby dinosaur fossils have been found in the stomachs of an ancient omnivorous mammal, *Repenomamus*, and we can safely assume that carnivorous dinosaurs occasionally returned the favor.

An alien scientist, observing our planet just before the Big One hit, would have had no reason to predict any great changes from then until now: dinosaurs would still be ruling the planet. Yet, in a clichéd "geolog-

Space artist Michael Carroll's wry view of the impact that sealed the fate of dinosaurs (except birds). COURTESY OF THE ARTIST ical blink of an eye," terrestrial dinosaurs were no more. Within a decade or so after the impact, they starved to death, photosynthesis having virtually ceased under the thick, long-lived global dust cloud kicked up by the asteroid. Plants died first, then creatures dependent on them for food, and then carnivores dependent on them, all the way up the food chain.

A few adaptable mammals, perhaps 20 species, somehow scraped and scavenged through the catastrophe, their small size (and therefore minimal calorie requirements and fast reproduction) giving them the edge over larger, hungrier animals. Anything bigger than a rat didn't make it. Once the atmosphere cleared and the climate settled down, photosynthesis restarted. Small, opportunistic mammals that survived the catastrophe—probably omnivores with an excellent sense of smell—emerged into a dinosaur-free world, and evolution did the rest, no doubt aided by the appearance of a new type of energy-rich plants: legumes. Within half a million years, a restored green Earth was home to mammals as large as sheep.

Today, mammals come in three groups, all of which have antecedents from that time: placentals, monotremes, and marsupials. While they have common features—warm-blooded metabolism, hair, and, of course, mammary glands for feeding their young (hence "mammals")—each group has its own characteristics. Monotremes, such as platypus and echnida, lay eggs, while marsupials, such as kangaroos, koalas, and opossums, give birth to small, frail babies that develop to term in their mothers' pouches. Placentals, by far the most successful—6,000 species of us—are characterized by giving birth to well-developed young, some of which are able to function effectively just minutes after birth (think of whales and wildebeests), while some take many years to be able to live independently (think of humans).

99

VULCAN, MISSING IN ACTION

Searching for Planet X

Vulcan was the Roman god of fire; he also was ugly and lame. The good news is that he somehow got to marry the goddess Venus, who was something of a catch. The bad news is that she was constantly unfaithful. Although sounding like a bit of a loser, Vulcan had the last laugh, giving us our word for one of the most potent of nature's forces: volcanoes. "Vulcan" crops up in other contexts too. For example, "vulcanization" is the process for hardening rubber. And (as all pre-TNG trekkies know), the imperturbable Mr. Spock, whose home planet is Vulcan, was the result of an improbable union between a Vulcan and a human.

My favorite Vulcan story, however, has to do with the search for a missing planet. More than 300 years ago, the grumpy genius Isaac Newton came up with his law of gravity, that the attractive force between any two objects depends on their masses and distance apart, thus explaining virtually all motion in our solar system. The law passed test after test, right up to when nineteenth-century astronomers began to take a really good look at the highly elliptical orbit of Mercury. The axis of Mercury's ellipse swings slowly around the sun in a motion known as "perihelion precession." About 200 years ago, astronomers measuring that precession found it wanting. While they could account for most of it by calculating gravitational effects of other planets, it was still a little more than that predicted by Newton's law. How come?

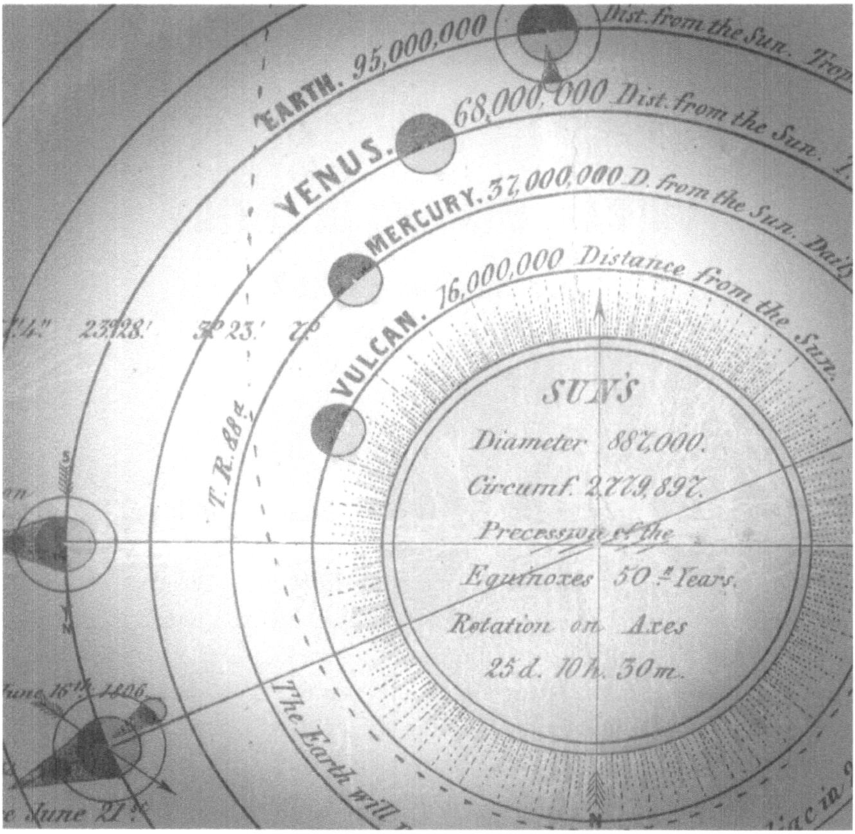

Detail from an 1846 lithograph of the solar system showing the planet Vulcan.
E. JONES AND G. W. NEWMAN, LIBRARY OF CONGRESS, PUBLIC DOMAIN

French astronomer Urbain Jean Joseph Le Verrier was the celestial whiz kid of his time, having correctly predicted the existence of Neptune from discrepancies in the orbit of Uranus. He proposed in 1859 that the discrepancy in Mercury's precession could be explained by a hitherto unseen planet that he called "Vulcan" orbiting between Mercury and the sun. The search began, and a few years later, Le Verrier naively put his trust in a single observation made by a French physician and amateur astronomer, Edmond Lescarbault, who claimed to have seen a new planet transit (pass across) the face of the sun. While Lescarbault was awarded the Légion d'honneur for his part, Le Verrier went on to calculate (to unbelievable precision) the orbital period of Vulcan: 19 days 17 hours at an inclination of 12 degrees 10 minutes.

Over the next half century, dozens of such transits were observed, all claiming to be proof of the existence of Vulcan. The trouble was that they disagreed with each other, and when the putative planet failed to show up during total solar eclipses, when it should have been readily observable, doubts crept in. It took a young German-born physicist reformulating Newton's "classical" law of gravity to unravel the conundrum. According to Albert Einstein's 1916 theory of general relativity, Newton's "spooky action at a distance" (where two masses are pulled toward each other by something like an invisible rubber band), what looks like attraction is actually a feature of the "warping" of an underlying ubiquitous matrix, spacetime. In Einstein's counterintuitive way of looking at it, *mass tells spacetime how to curve, while the curvature of spacetime tells mass how to move.*

When astronomers allowed for the curvature of spacetime caused by the sun's mass, the discrepancy disappeared. In fact, this was the first empirical test of Einstein's theory, which it passed with flying colors. So the raison d'être for Vulcan's existence—not the Roman god of fire or Spock's home but a planet once thought to orbit very close to the sun—went away. Too bad. "Vulcan" sounds so right for the name of a fiery, sun-blasted world.

100

ARE WE THE WORLD?

The problem with hubris

According to the Roman writer Ovid, the hunter Narcissus was handsome as all get-out, causing the mountain nymph Echo to fall in love with him. When he rejected her advances (the fool!), she pined away until all that was left was a faint sound, hence our word *echo*. This led to Aphrodite, goddess of revenge, to punish him by luring him to a pool of water, where he fell in love with his own reflection. Unable to consummate this love, his passion led to his demise, turning him into a white and gold flower, the narcissus.

Many writers, past and present, have glommed onto this sad story of what we now call "narcissism" as a way of explaining humankind's naive way of seeing our species as super-important: from *geocentrism*, in which everything revolves around Earth; to *religions*, in which God or the gods are obsessed by our every move (perhaps having first created us in their image); to *consciousness*, which, according to some, is the fundamental stuff of reality. (As far as we know, consciousness is a phenomenon experienced by one out of millions of species on one out of billions of planets. Hardly the stuff of universal truisms.)

I was born a narcissist. I am the epicenter of reality. Fortunately for my hubris, at some point in my childhood, I realized that this was true of everyone; we *all* seem to be in the middle of everything. Similarly with our conception of Earth's place in the cosmos. Yet the inhabitants of *any*

Geocentric model according to Claudius Ptolemy (AD 100–170), from Peter Apian's Cosmographia, 1524. PUBLIC DOMAIN

planet would think themselves to be at the center of their solar system! Somehow, that obvious fact escaped ancient and medieval inquirers who instead accepted the geocentric view of Ptolemy, who, writing nearly 2,000 years ago, said that Earth is at the center and that the planets, the sun, and other stars revolve around us. It took the combined efforts of several brave observers and theorists, notably Copernicus, Kepler, and Galileo, to displace Earth from this exalted cosmic throne.

More humbling was to follow. Only a century ago, conventional wisdom held that the Milky Way was the entirety of the universe until Heber Curtis and Edward Hubble, working with photographs taken by the then-new 100-inch Mount Wilson telescope, showed that our galaxy was one of many. And so our self-importance took one hit after another. We're not the center of the solar system, the solar system isn't the center of the Milky Way, and the Milky Way isn't the center of the universe. So much for geocentrism.

More recently, science writer John Horgan coined the term *neo-geocentrism* in describing the wacky view that *consciousness* is central to, well, everything, pervading the entire universe—perhaps even responsible for life itself. ("The origin of life forms is the differentiation of pure consciousness into multiple forms of life," according to New Ager Deepak Chopra.) Some well-known scientists have jumped on the "consciousness is fundamental" bandwagon, including psychologist Donald Hoffman, neuroscientist Christof Koch, and physicist Max Tegmark. Their views echo those of old-time panpsychists, who believed that *everything* was conscious!

Beyond consciousness, another example of neo-geocentrism is the anthropic principle, the notion that the universe must be as it appears to us, or else we wouldn't be here to observe it. Change, just a tad, one of the fundamental physical constants—the ratio of the electromagnetic force to the gravitational force, for example—and you wouldn't have a universe with stars, planets, and observers. Cosmologists Stephen Hawking and Brian Greene signed on to this "it's all about us" point of view. And if that's not spooky enough for you, how about late physicist John Wheeler's "participatory universe," which asks, "Does the universe even *exist* if there's no one to observe it?"

I'm with Horgan. Such narcissistic approaches to understanding the universe are modern, anthropocentric versions of ancient superstitions, in his words, "the projection of our fears and desires, our yearning to matter." Isn't it wonder enough that our three-pound brains can conceive of a 14-billion-year-old universe rather than trying to make us—and our consciousness—essential to its very existence?

Maybe it's time to get over ourselves.

101

Z IS FOR ZODIAC

I am Virgo, hear me roar

I'm a Leo, honest. I was born on September 6, 1942, when the sun was firmly in the constellation Leo. Another couple of degrees to the west, and it would have eclipsed Regulus, Leo's brightest star. Yet my wife (born August 2) happily—smugly even—points out that every newspaper horoscope classifies me as a Virgo. What's going on? Why isn't my horoscope sign my sun sign? How come I'm cheated out of my leonine birthright?

Around 600 BC, the ancient Babylonians divided the ecliptic—the sun's annual path through the sky relative to the background stars—into 12 parts, each one corresponding to a constellation, making up the zodiac (that is, relating to animals, cognate with *zoo*). So, for example, if you were born during the 30-day period when the sun was "in" Capricorn, your sun sign was Capricornus. At the same time, they came up with the idea of the *vernal equinox*. Imagine two vast circles in space, one an extension of Earth's equator (the *celestial equator*) and the other the plane of Earth's orbit around the sun (the *ecliptic*). The vernal equinox is one of the two annual points in the sky where the two circles appear to intersect.

The Babylonians' calendar year started with the vernal (Spring) equinox, which in their era had the sun in Aries. Five hundred years later, Greek math whiz Hipparchus (190–120 BC) referred to old Sumerian records to show that the vernal equinox appeared to be moving along the ecliptic. Today, the vernal equinox is in Pisces, the next constellation

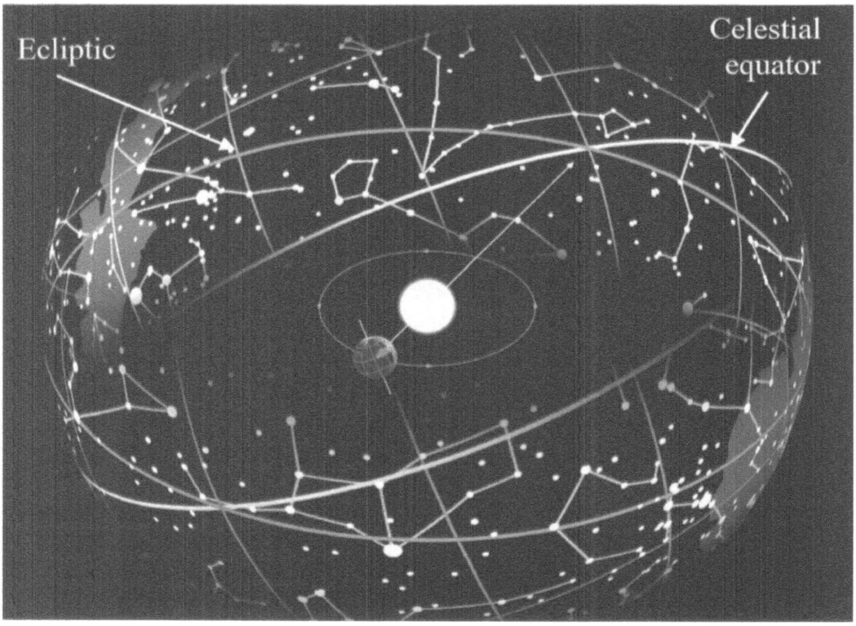

Seen from Earth, the sun moves along the ecliptic. The intersection of the ecliptic and celestial equator defines the vernal equinox, now "in" Pisces. For the ancient Babylonians, it was in Aries. TAU'OLUNGA/GNU FREE DOCUMENTATION LICENSE

over from Aries, meaning that someone born when the sun is "in" Leo is (unfairly!) branded a Virgo.

Blame the precession of Earth's axis for the discrepancy. Like a spinning top, Earth's spin precedes, or wobbles, as its axis of rotation traces out a cone. For that, you can blame our sun, which rotates about once every 27 days (it varies with latitude) relative to the background stars. The sun's spin exerts a torque on the spinning Earth, causing it to wobble. One full "wobble" takes 26,000 years, so we'll be back to where the Babylonians started out in another 23,000-odd years.

Astrologers know all this, of course, which is why there are two basic schools: tropical astrology and sidereal astrology. The basis for *sidereal astrology* is to note the actual constellation in which the sun is located at the moment of birth (making me a Leo). However, the more popular *tropical astrology* uses a 30-degree sector of the zodiac (according to Greek astronomer-geographer Ptolemy, AD 100–170), ignoring the actual positions of the sun, planets, and stars.

So, sidereally, I'm a roaring Leo (confident, generous, loyal), while tropically, I'm one of those analytical, fussy, skeptical Virgos. That can't be right. Me, *skeptical?*

ACKNOWLEDGMENTS

Amazon to Zodiac has a long and rich history. Almost all the chapters here started life as "Field Notes" columns in Humboldt County, California's *North Coast Journal*, a weekly "politics, people and art" paper. For nearly two decades, Field Notes has been a home for an eclectic miscellany of essays on science, nature, local history, health, and more. This book is an edited, updated, and expanded collection of what I consider to be the best 101 topics from a basket of about 600 columns.

Many people helped steer me to this compendium, in particular two hard-nosed editors at the *Journal*, originally Hank Sims, now Jen Fumiko-Cahill. (If I'm bucking the *Chicago Manual of Style*, it'd better be for a very good reason!) Thank you both for the latitude and encouragement you've given me over the years.

The two essential qualities for writing about science are boundless curiosity matched with boundless skepticism. The first I was happily born with, while the second came more slowly, with the help of many teachers and collaborators over the years. In particular, I'm indebted to several kind people no longer with us, including Martin Gardner, Frank Drake and Philip Morrison. Also, thanks to still-on-the-planet Chris McKay, Michael Carroll, Scott Sandborn and Paul Boynton.

More recently, I'm grateful to Dave Fitzgerald, Victoria Andrews, Jerry Rohde, Bronco McCoy, and dozens, if not hundreds, of scientists, researchers, artists, and curators who have been kind enough to share their knowledge and free use of graphics.

The patient folks at Prometheus Books have been great to deal with. Thanks to Jake Bonar and Alexis Vaughn, not forgetting the late founder of the company, humanist Paul Kurtz.

ACKNOWLEDGMENTS

The writing of a 1,000-word science essay usually begins with a single entry in Wikipedia. The best thing about Wikipedia is that it always cites sources, which inevitably lead me into deep tunnels of fact-finding and analysis. Thank you editors, and especially Jimmy Wales, whose vision a quarter of a century ago led to the largest reference work in history.

My partner of over 52 years – wife, editor, and writer herself—is the one who suggested, half a century ago, that I write my first book. Thank you, Louisa Rogers.

INDEX

Note: Page references for figures are italicized

INDEX

INDEX